Lecture Notes on Theoretical Mechanics

Jianlin Liu

Lecture Notes on Theoretical Mechanics

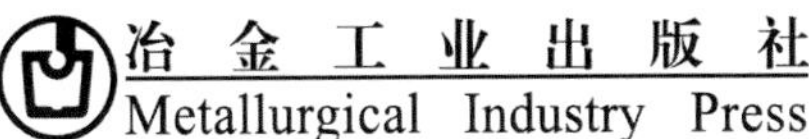

Jianlin Liu
Department of Engineering Mechanics
College of Pipeline and Civil Engineering
China University of Petroleum (East China)
Qingdao, Shandong, China

ISBN 978-981-13-8037-2 ISBN 978-981-13-8035-8 (eBook)
https://doi.org/10.1007/978-981-13-8035-8

Jointly published with Metallurgical Industry Press, Beijing, China
The print edition is not for sale in China Mainland. Customers from China Mainland please order the print book from: Metallurgical Industry Press.
ISBN of the China Mainland edition: 978-7-5024-8011-0

This Springer imprint is published by the registered company Springer Nature Singapore Pte Ltd.
The registered company address is: 152 Beach Road, #21-01/04 Gateway East, Singapore 189721, Singapore

Preface

Abstract In this chapter, we give a simple introduction on the history of mechanics, especially on applied mechanics. The content of the course "*Theoretical Mechanics*" is also simply introduced.

Keywords Theoretical Mechanics • History of mechanics • Applied mechanics

As we all know, the word "mechanics" is quite similar to the word "mechanical", but we can see that it has a special letter "s", which is the abbreviation of "science". This indicates that mechanics is an independent subject. In fact, mechanics was closely related with mechanical engineering in the ancient time, meaning "manufacture" and "design"; but it also has the theoretical characteristics of rigorous logistic derivation. Then a critical question is "What's mechanics?" Generally speaking, mechanics is a subject to investigate the laws of force and motion. Engineering mechanics is aimed to solve engineering design problems in engineering areas, such as aircrafts in aerospace, automobiles, civil engineering, petroleum engineering, material preparation, mechanical engineering, and architecture engineering.

The origins of mechanics are lost in antiquity. There were numerous famous scholars who made contributions to engineering mechanics. For example, Aristotle (384 BC–322 BC), a scholar of encyclopedia, gave the first thinking to motion, velocity, and force. The Great scientist Archimedes (287 BC–212 BC), an encyclopedic scholar, proposed a lot of physical laws and mechanical equipments on mechanics. For instance, the famous Archimedes law of buoyant force, Archimedes lever rule, and calculation methods of volume, area, and center of mass on geometric bodies were all given by Archimedes. Actually, he was regarded as one of the four greatest mathematicians in history, for he gave the definition of limitation in mathematics. Moreover, Simon Stevin (1548–1620) proposed the principle of virtual work, and he was viewed as the father of Statics. He also gave the parallelogram law on force summation. In addition, the mathematician Johannes Kepler (1571–1630) proposed the three motion laws of planet. Another encyclopedic scientist is Leonardo da Vinci (1452–1519), who drew a lot of aircrafts (as shown in Fig. 1), automobiles, bicycles, and some other mechanical equipment in his drafts, which were not published in his era. Da Vinci also explored the strength of materials, and pointed out the importance of flaws in materials. Certainly, he was also a great painter with a lot of famous works, such as *Mona Lisa* and *the Last Supper*. Shortly after da Vinci, the great mechanics scholar Galileo Galilei (1564–1642) was born. He was viewed as the father of classical mechanics and experiments, who introduced the new concept of acceleration, which depicts the variation ratio of the velocity. As a consequence, he was thought of as the father of dynamics. Galileo was also one of the predecessors who explored the secret of material strength. He performed the first experiment all over the world to examine the strength of a cantilever, which was schematized in Fig. 2. Based upon the experimental results, he proposed the so-called "first strength theory", still adopted in the textbook of today. In what follows, several French scientists were eager to advance the development of mechanics. Among others, René Descartes (1596–1650), who was also a philosopher, proposed the concept of "momentum" and "theorem of momentum". One of his important contributions is the introduction of Cartesian coordinate system, which laid the foundation of analytical geometry. He was also concerned with optics, universe evolution, and biological system of human body. One of his peers, Blaise Pascal (1623–1662), who was also a genius, proposed the law of Pascal pressure. Moreover, he was the grandfather of computer and probability theory in mathematics. Especially, he was a famous philosopher and a prose writer. Nearly at the same time, a neglected scientist Pierre-Louis Moreau de Maupertuis (1698–1759) proposed the principle of least action, which was used to analyze the transmission of light, and then extended to mechanics. Following these giants, an epochal scholar, Isaac Newton (1643–1727) (Fig. 3) was born. He concluded the three Newton's laws of motion, Newton's law of universal gravity, Newton optics, Calculus, etc. He was regarded as one of the greatest scientists in the human history. He was so famous in the scientific world that he was awarded many honors before he died. In fact, this great giant was fighting with his competitors all through his life, among which including Robert Hooke (1635–1703), Gottfried Wilhelm Leibniz (1646–1716), and Christiaan Huygens (1629–1695). The elaborate Hooke's law was named by Robert Hooke, but it has been verified that it was first proposed by a Chinese scholar Xuan Zheng (郑玄, 127–200) in the Han Dynasty. Leibniz was another founder of calculus as the biggest competitor of Newton, and he also proposed the theorem of kinetic energy. He designed the first machine to perform the multiplication operation, which had a great influence on computer. He was so learned that he knew law, diplomacy, metallurgy, chemistry, physics, and philosophy. After Newton, a lot of familiar names appeared, including Leonhard Euler (1707–1783), J. L. Lagrange (1736–1813), J. C. F. Gauss (1777–1855), Augustin L. Cauchy (1789–1857), P. S. Laplace (1749–1827), Simeon D. Poisson (1781–1840), William R. Hamilton (1805–1865), Claude-Louis Navier (1785–1836), Saint Venant (1797–1886), George G. Stokes (1819–1903), etc. These scholars further laid the foundations of the mechanics building. For example, Euler proposed the stability theory of slender structures and motion equation of rigid bodies (as shown in Fig. 4), Lagrange set up the Lagrange equation, which paved a new way to analytical mechanics, and Gauss proposed the principle of least constraint and advanced differential geometry.

Some other theorems and laws are also related with these names, such as Bernoulli equation, Cauchy stress, Cauchy's theorem, Poisson's ratio, Poisson's distribution, Laplace equation, Hamilton principle, Navier–Stokes equation, and principle of Saint Venant. In this period, the fundamental theory of continuum mechanics was built based upon these pioneering contributions.

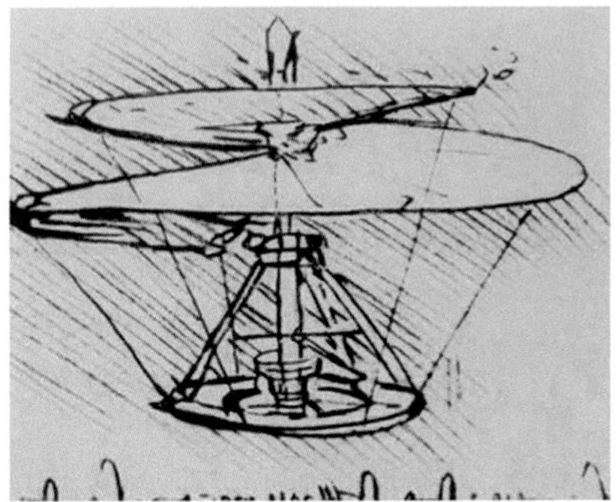

Fig. 1 Aircraft designed by da Vinci

Fig. 2 The first cantilever experiment by Galileo

The work mentioned above was actually the elementary theory of applied mechanics. During World War I, a lot of scientists had some new understandings on the importance of mechanics to weapons and industries. The famous mathematician Felix Klein (1849–1925), who was the mathematics leader in Göttingen University, proposed the spirit of applying mathematics to solve engineering problems. He made the applied mechanics scholar Ludwig Prandtl (1875–1953) move to Göttingen to design one of the earliest wind tunnels in the world. Prandtl

Fig. 3 Newton's apple

Fig. 4 Rotation of a gyroscope

established a research institute in Göttingen, aiming to investigate applied mechanics issues in engineering. His most famous theory, the boundary layer theory made the Navier–Stokes equation simplified, which was the basic theory in the aircraft design. Moreover, Prandtl cultivated a lot of scholars in mechanics, such as S. Timoshenko (1878–1971), who wrote a series of popular textbooks which were still adopted nowadays; Theodore von Kármán (1881–1963), who later went to Caltech to build another center of applied mechanics; William Prager (1903–1980), who was an applied mathematician in Brown University. Prandtl had only one female student, Shi-jia Lu (1911–1986), who came from China and later also became a famous scientist. Altogether, this school inherited from Prandtl was called Göttingen applied mechanics school. It should be mentioned that China's mechanics career originated from Göttingen school. The founders of Chinese mechanics career are Hsueshen Tsien (1911–2009) (Fig. 5), Peiyuan Zhou (1902–1993), Weichang Qian (1912–2010), and Yonghuai Guo (1909–1968). Peiyuan Zhou was the student of Werner K. Heisenberg (1901–1976) and Albert Einstein

(1879–1955), whose majors are turbulence and theory of relativity. Except Prof. Zhou, the other three scholars were all students of Prof. von Kármán in Caltech. When they returned to China, they led the rocket, satellite, and atomic bomb careers, and also laid the foundations of mechanics education and research in China. These four scholars were also the founders of the Chinese Society of Theoretical and Applied Mechanics.

Fig. 5 Prandtl, Hsueshen Tsien and von Kármán

It can be seen that mechanics is not far away from us; in fact, it exists widely in nature, industry, and our daily life. We always face so many questions on the secret of nature: Why does the bird fly in the sky freely, and why is it so difficult for an airplane to fly? Why can the fish swim fluently in water, and why is it so hard for a ship to swim? Why can some creatures jump on the liquid surface and why is it impossible for a human to move on water? How to design a bridge or a building in prevention of wind, snow, and earthquake? If there is a crack in a machine or a ship, can it still work? If can, how long will it last? How to protect the passengers in a car when crashed? How to design a structure with least materials and lowest cost? All these questions are related with mechanics. In the current education system of China, the class of Engineering Mechanics includes two portions, namely, Theoretical Mechanics and Mechanics of Materials (or Strength). Mechanics of materials is aimed to investigate the deformation and stress of structures, which will be introduced after the current task. We mainly concentrate on Theoretical Mechanics in this course, which deals with the force equilibrium and motion laws on rigid bodies.

A rigid body is only a perfect model, which is assumed to be without experiencing deformation under the action of external forces. In other words, the distance of two arbitrary points in the rigid body never changes, and the shape of the whole rigid body does not alter. In the practical world, it is impossible to find a real rigid body. However, when the deformation is not important to the property of the problem, it can be neglected and the object can be thought of as a rigid body.

The class of Theoretical Mechanics embraces three sections, namely, *Statics*, *Kinematics*, and *Kinetics*. The latter two sections are often named as *Dynamics*. Due to so colorful contents of theoretical mechanics, it is not easy to introduce all the knowledge in this short course, including only 48 lessons. Another fact we should face is that not all the students are in the same level. Therefore, we will give the fundamental knowledge in the second chapter, in order to make the students recall the past equations and formulas. We then simplify most knowledge of Theoretical Mechanics, but it still includes three sections, i.e., Statics, Kinematics, and Kinetics. From this textbook, the author expects the students to master the skill of force analysis, writing out the equilibrium equations, velocity analysis of rigid body, and using the fundamental laws in dynamics.

In Statics, we only focus on the force analysis, force group simplification, and equilibrium conditions of the rigid body. In Kinematics, we analyze the motion laws of the particle and rigid body from the viewpoint of geometry, not considering forces. In the last portion, i.e., Kinetics, we investigate the relations between the motion and external force by introducing the theorems in dynamics, i.e., Newton's second law, theorem of momentum, theorem of angular momentum, and theorem of kinetic energy.

We thank the useful comments from Dr. Dongying Liu, Dr. Jing Li, Miss Caixia Hu, Miss Xian Han, and the graduate students with the names of Jing Sun, Pingcheng Zuo, Gaofeng Cao, Shanpeng Li, and Yulong Gong. Although we have tried our best to write this lecture note, there may still be some errors. We are expecting the comments from all the readers, and then we hope to give a better version of this lecture note in the near future.

With my best regards,

Qingdao, China Jianlin Liu
January 2019

Contents

Chapter 1
Preliminary Knowledge

Abstract In this chapter, we give some fundamental knowledge to learn *Theoretical Mechanics*, which is very necessary. The content includes three sections, i.e., the trigonometric function, concept of distance, scalar, and vector.

Keywords Trigonometric function · Distance · Scalar · Vector

1.1 Trigonometric Function

We have already learned the trigonometric functions in the middle school stage, and here let's first have a brief overview about these kernel definitions. We can see that in the forthcoming content of this curriculum, it is quite necessary to master these definitions and calculations very fluently.

We consider a right triangle with an acute triangle, where its three sides are a, b, and c, as schematized in Fig. 1.1. The angle between b and c is θ.

According to Fig. 1.1, the most common trigonometric functions and correlated relations are formulated as

$$\begin{aligned} \sin\theta &= \tfrac{1}{\csc\theta} = \tfrac{a}{c}, \\ \cos\theta &= \tfrac{1}{\sec\theta} = \tfrac{b}{c}, \\ \tan\theta &= \tfrac{a}{b} = \tfrac{\sin\theta}{\cos\theta} = \tfrac{1}{\cot\theta}, \\ \sin^2\theta &+ \cos^2\theta = 1. \end{aligned}$$

In practice, if the side length c or b is already known, we normally use the following formats:

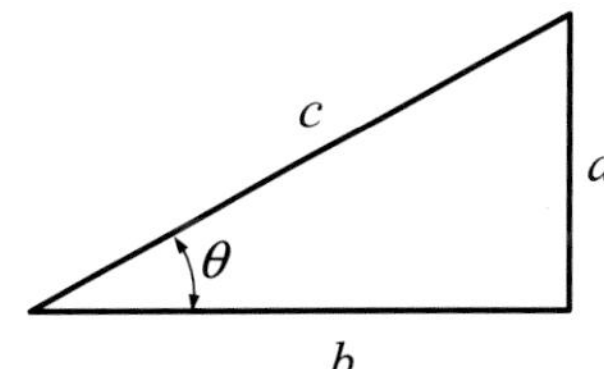

Fig. 1.1 A right triangle

J. Liu, *Lecture Notes on Theoretical Mechanics*,
https://doi.org/10.1007/978-981-13-8035-8_1

$$
\begin{aligned}
a &= c \sin\theta, \\
b &= c \cos\theta, \\
a &= b \tan\theta.
\end{aligned}
$$

Furthermore, we should remember the actual values of the trigonometric functions about the special angles, such as 0, 30°, 45°, 60°, 90°, 120°, 150°, and 180°. The detailed relations are listed as follows:

$$
\begin{aligned}
\sin 30^\circ = \cos 60^\circ = \sin 150^\circ &= \tfrac{1}{2}, \\
\cos 30^\circ = \sin 60^\circ = \sin 120^\circ &= \tfrac{\sqrt{3}}{2}, \\
\sin 45^\circ = \cos 45^\circ = \sin 135^\circ &= \tfrac{\sqrt{2}}{2}, \\
\sin 0^\circ = \cos 90^\circ = \sin 180^\circ &= 0, \\
\cos 0^\circ = \sin 90^\circ &= 1, \\
\tan 30^\circ = \cot 60^\circ &= \tfrac{\sqrt{3}}{3}, \\
\cot 30^\circ = \tan 60^\circ &= \sqrt{3}, \\
\tan 0^\circ = \cot 90^\circ &= 0, \\
\tan 45^\circ = \cot 45^\circ &= 1, \\
\tan 90^\circ = \cot 0^\circ &= \infty, \\
\cos 150^\circ &= -\tfrac{\sqrt{3}}{2}, \\
\tan 150^\circ &= -\tfrac{\sqrt{3}}{3}, \\
\cos 120^\circ &= -\tfrac{1}{2}, \\
\tan 120^\circ &= -\sqrt{3}, \\
\cos 180^\circ &= -1.
\end{aligned}
$$

Besides these, we had to better remember the following right triangle with three sides being round numbers. For example, if the two smaller sides are 3 and 4, then the third side should be 5. Similarly, if the former two are 6 and 8, then the third side is 10. There are also related round numbers, such as 5, 12 and 13; 7, 24, and 25; 9, 40, and 41; 11, 60, and 61; 13, 84, and 85. Again, for the (3, 4, and 5) triangle, if

$$
\begin{aligned}
\sin\theta &= \frac{3}{5} = 0.6, \\
\cos\theta &= \frac{4}{5} = 0.8,
\end{aligned}
$$

we have $\theta = 37^\circ$.

1.2 Concept of Distance

In the following chapters, we will often meet the concept of "distance", so it is necessary to recall these definitions.

As shown in Fig. 1.2, it is seen that there are many paths linking point *A* and *B*, where the straight segment *AB* is the shortest route. Therefore, the distance between the two points *A* and *B* is just the length of the straight segment *AB*.

Next, as shown in Fig. 1.3, the length *AB* is the distance of point *A* to the horizontal line, where *AB* is perpendicular to the horizontal line. This also means that *AB* is the distance between the point and the line.

Similarly, the distance between two parallel lines is shown in Fig. 1.4. We can select any point *A* in the first line, then draw a line perpendicular to the second line, and the cross point is *B*. As a result, *AB* is the distance between these two parallel lines.

We then look at the distance between a point and one plane. As shown in Fig. 1.5, line *AB* is normal to the plane, and point *B* is the cross point. Therefore, the length of *AB* is the distance between point *A* and the plane.

As shown in Fig. 1.6, one line is parallel to a plane. Then we select an arbitrary point *A* on the line, and draw another line vertical to the plane, where point *B* is the cross point. Then *AB* is the distance between the line and the plane.

As shown in Fig. 1.7, two planes are parallel to each other. If we select a point *A* in the upper plane, then we can get the distance between point *A* and the bottom plane. Thus, *AB* is the distance between the two parallel planes.

At last, let's look at the distance between two noncoplanar lines. As shown in Fig. 1.8, there are two lines in different planes, and they are not parallel or crossed. We can first make a line in the bottom plane, which is parallel to the upper line. The new line is crossed with the second line at point *B*. Then line *AB* is perpendicular to the bottom plane, which is the distance between the original noncoplanar lines.

As a typical example, let us look at the cuboid *ABCD-EFGH*, as shown in Fig. 1.9. *AB* and *FG* are noncoplanar, and the distance between them is *BF*, *AE*, *CG*, or *DH*.

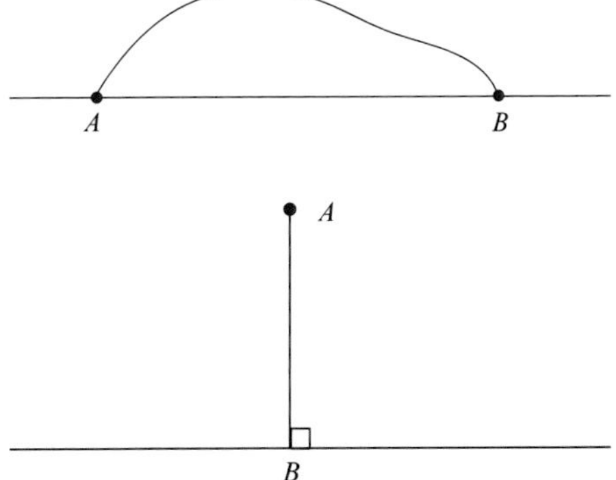

Fig. 1.2 Distance of two points

Fig. 1.3 Distance of one point to one line

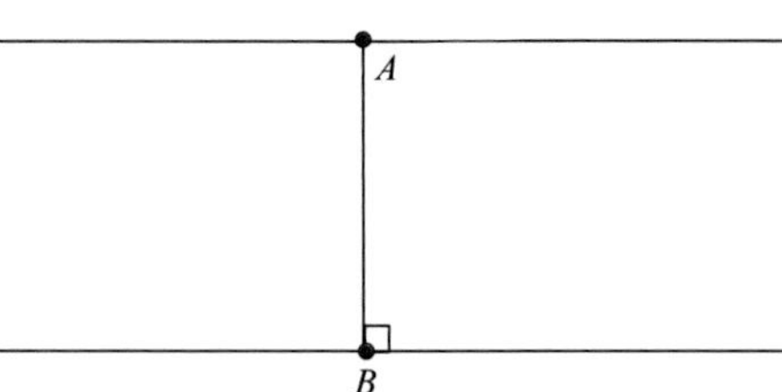

Fig. 1.4 Distance of two parallel lines

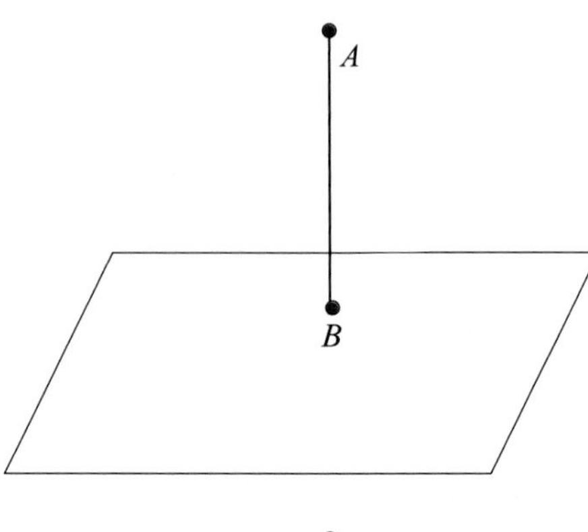

Fig. 1.5 Distance of one point to one plane

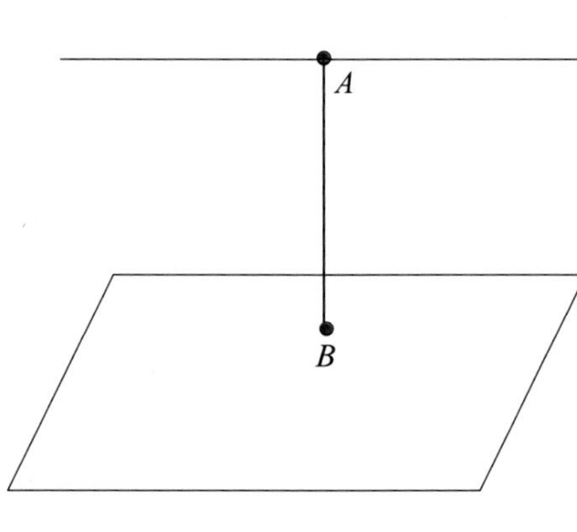

Fig. 1.6 Distance of one line to a parallel plane

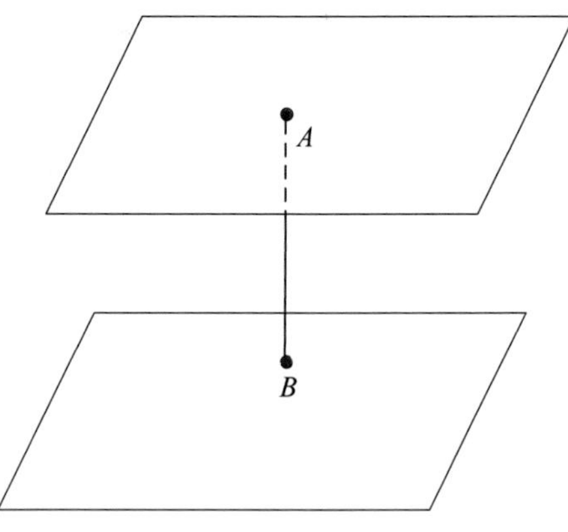

Fig. 1.7 Distance between two parallel planes

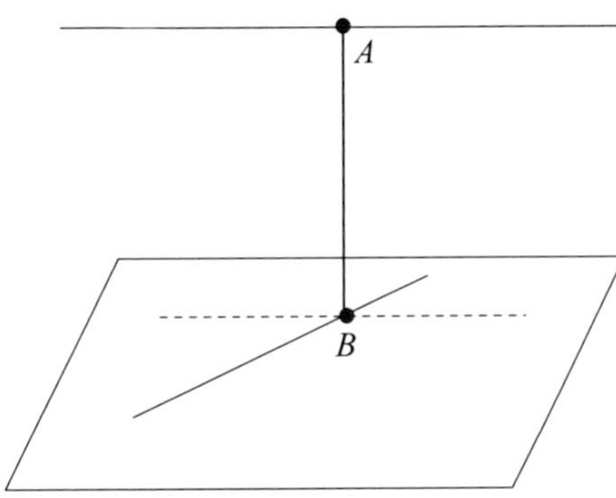

Fig. 1.8 Distance of two noncoplanar lines

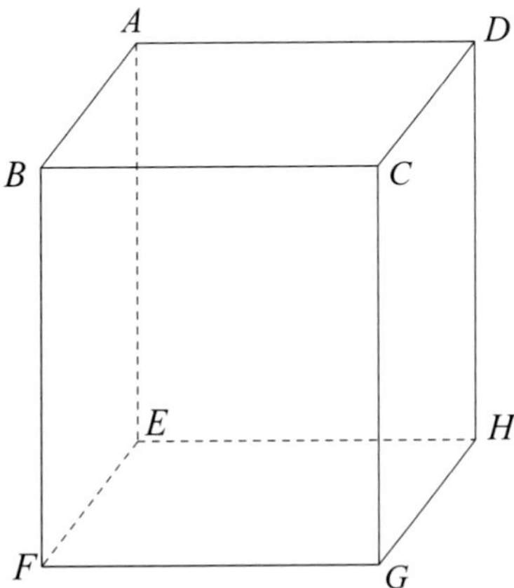

Fig. 1.9 Cuboid

BC and *DH* are also noncoplanar, and their distance is *AB*, *CD*, *GH*, or *EF*. It will be seen that it is convenient to grasp the distance between two noncoplanar lines later.

1.3 Scalar and Vector

Most of the physical quantities can be divided into two types, i.e., scalars and vectors. A scalar only has magnitude and no direction, such as time t, temperature T, mass m, volume V, density ρ, and energy E. However, a vector is more complex, for it has both magnitude and direction, such as displacement, velocity, acceleration, gravity, force, and moment. We will later see that "vector" is really an important and fundamental concept in mechanics, so it is essential to master the elementary of this quantity.

A general vector can be defined as

$$\boldsymbol{a} = a\boldsymbol{n},$$

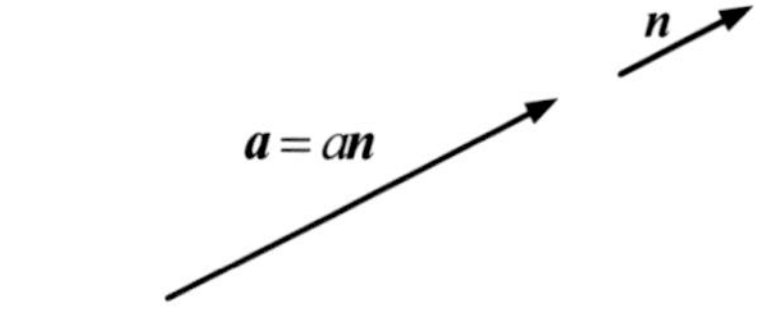

Fig. 1.10 Vector

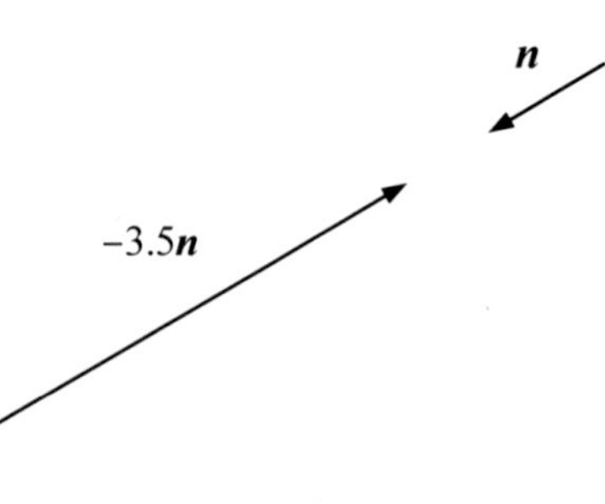

Fig. 1.11 Example of a vector

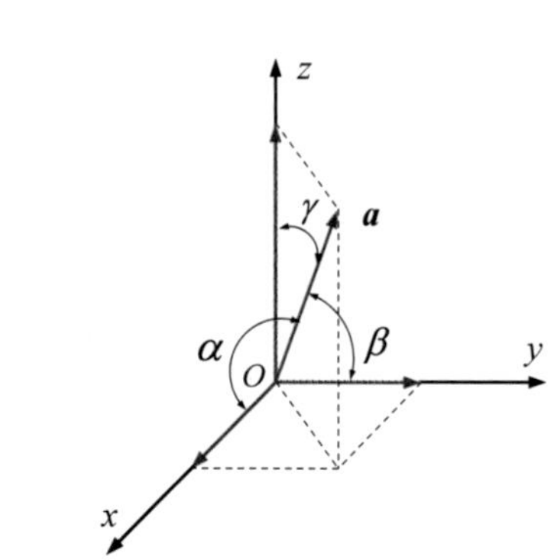

Fig. 1.12 A vector in the Cartesian coordinate system

where a is the magnitude of the vector $\boldsymbol{a}$ and $\boldsymbol{n}$ is a unit vector with the magnitude of 1, which is of the same direction of vector $\boldsymbol{a}$, as shown in Fig. 1.10. The norm or the absolute value of $\boldsymbol{a}$ is

$$a = |\boldsymbol{a}|.$$

For example, the vector $\boldsymbol{a} = 3.5\boldsymbol{n}$ means the magnitude of the vector is 3.5, and its direction is along that of the unit vector $\boldsymbol{n}$. The vector $\boldsymbol{a} = -3.5\boldsymbol{n}$ means the direction of $\boldsymbol{a}$ is opposite to that of the unit vector $\boldsymbol{n}$. The vector can be schematized in Fig. 1.11.

In fact, to analyze the properties of a vector, it is natural to put it in the Cartesian coordinate system o-xyz, as shown in Fig. 1.12. The three axes x, y, and z of the coordinate system obey the right-handed screw rule.

A vector is not convenient to calculate, and therefore we often consider the projections of the vector. In the Cartesian coordinate system of Fig. 1.12, the vector $\boldsymbol{a}$ can be expressed as

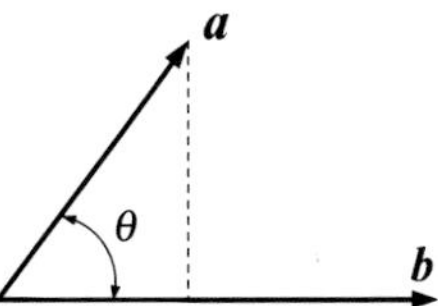

Fig. 1.13 Dot product

$$\boldsymbol{a} = a_x\boldsymbol{i} + a_y\boldsymbol{j} + a_z\boldsymbol{k} = \left(a_x, a_y, a_z\right),$$

where a_x, a_y, and a_z are the three components of the vector, and $\boldsymbol{i}, \boldsymbol{j}$, and $\boldsymbol{k}$ are the unit vectors of axis x, y, and z, respectively. That means a_x, a_y, and a_z represent the three projections of the vector $\boldsymbol{a}$ along the three axes. The magnitude of the vector $\boldsymbol{a}$ is expressed as

$$|\boldsymbol{a}| = \sqrt{a_x^2 + a_y^2 + a_z^2}.$$

For a vector in the o-xy plane, it can be written as

$$\boldsymbol{a} = a_x\boldsymbol{i} + a_y\boldsymbol{j}.$$

We then define the dot product of two vectors as

$$\boldsymbol{a} \cdot \boldsymbol{b} = ab\cos\theta,$$

where θ is the angle between $\boldsymbol{a}$ and $\boldsymbol{b}$, which is schematized in Fig. 1.13. Then we have

$$\boldsymbol{i} \cdot \boldsymbol{i} = 1, \quad \boldsymbol{i} \cdot \boldsymbol{j} = 0, \quad \boldsymbol{i} \cdot \boldsymbol{k} = 0,$$

$$\boldsymbol{j} \cdot \boldsymbol{j} = 1, \quad \boldsymbol{j} \cdot \boldsymbol{k} = 0, \quad \boldsymbol{k} \cdot \boldsymbol{k} = 1.$$

From the geometric viewpoint, $a\cos\theta$ is the projection of the vector $\boldsymbol{a}$, which is of the same direction as that of the vector $\boldsymbol{b}$. Inverting the components of the vectors the dot product can be written as

$$\boldsymbol{a} \cdot \boldsymbol{b} = a_x b_x + a_y b_y + a_z b_z.$$

Maybe one typical example of the dot product is the work, which equals the product of the equivalent force and distance. Following this definition, one has

$$\begin{aligned} a_x &= \boldsymbol{a} \cdot \boldsymbol{i} = a\cos\alpha, \\ a_y &= \boldsymbol{a} \cdot \boldsymbol{j} = a\cos\beta, \\ a_x &= \boldsymbol{a} \cdot \boldsymbol{k} = a\cos\gamma, \end{aligned}$$

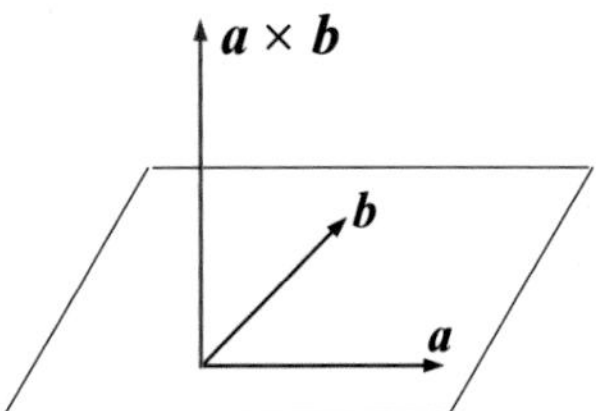

Fig. 1.14 Cross product

where α, β, and γ are the angles between the vector and axis x, y, and z, respectively, which are displayed in Fig. 1.12. Another common format of these relations is

$$\cos\alpha = \frac{a_x}{a},$$
$$\cos\beta = \frac{a_y}{a},$$
$$\cos\gamma = \frac{a_z}{a}.$$

Following the above definitions, the vector $\boldsymbol{a}$ can be rewritten as

$$\boldsymbol{a} = a(\cos\alpha\boldsymbol{i} + \cos\beta\boldsymbol{j} + \cos\gamma\boldsymbol{k}) = a\boldsymbol{n},$$

where the unit vector

$$\boldsymbol{n} = \cos\alpha\boldsymbol{i} + \cos\beta\boldsymbol{j} + \cos\gamma\boldsymbol{k}$$

and

$$\cos^2\alpha + \cos^2\beta + \cos^2\gamma = 1.$$

Another operation about two vectors is the cross product, which is defined by the determinant as follows

$$\boldsymbol{a}\times\boldsymbol{b} = -\boldsymbol{b}\times\boldsymbol{a} = \begin{vmatrix} \boldsymbol{i} & \boldsymbol{j} & \boldsymbol{k} \\ a_x & a_y & a_z \\ b_x & b_y & b_z \end{vmatrix} = \left(a_x b_y - a_y b_x\right)\boldsymbol{k} + \left(a_y b_z - a_z b_y\right)\boldsymbol{i} + (a_z b_x - b_z a_x)\boldsymbol{j}.$$

This kind of operation introduces a new vector, whose direction is perpendicular to the coplane determined by the two vectors $\boldsymbol{a}$ and $\boldsymbol{b}$, which is schematized in Fig. 1.14. The direction of the cross product of $\boldsymbol{a}$ and $\boldsymbol{b}$ can also be determined by the right-handed screw rule. The rolling direction is from $\boldsymbol{a}$ to $\boldsymbol{b}$, and then the direction of the cross product is the normal direction of the coplane.

The other operations about vectors are

$$\boldsymbol{c} = \boldsymbol{a} \pm \boldsymbol{b} \Leftrightarrow \begin{cases} c_x = a_x \pm b_x \\ c_y = a_y \pm b_y \\ c_z = a_z \pm b_z \end{cases},$$

$$\boldsymbol{a} + \boldsymbol{b} = \boldsymbol{b} + \boldsymbol{a},$$

$$\boldsymbol{a} - \boldsymbol{b} = \boldsymbol{a} + (-\boldsymbol{b}),$$

$$\boldsymbol{a} + \boldsymbol{b} + \boldsymbol{c} = \boldsymbol{a} + (\boldsymbol{b} + \boldsymbol{c}),$$

$$\boldsymbol{c} = \lambda \boldsymbol{a} \Leftrightarrow \begin{cases} c_x = \lambda a_x \\ c_y = \lambda a_y \\ c_z = \lambda a_z \end{cases},$$

where λ is a scalar constant. The addition and subtraction of vectors obey the parallelogram law, which will be further discussed in the next chapter.

Example 1 Two vectors $\boldsymbol{A} = 3\boldsymbol{i} - 7\boldsymbol{j} + 11\,\boldsymbol{k}$, $\boldsymbol{B} = \boldsymbol{i} - 2\boldsymbol{j} + 6\,\boldsymbol{k}$.

Answer:

$$\boldsymbol{A} \cdot \boldsymbol{B} = 3 \times 1 + 7 \times 2 + 11 \times 6 = 83 = \boldsymbol{B} \cdot \boldsymbol{A}$$

$$\boldsymbol{A} - 2\boldsymbol{B} = 3\boldsymbol{i} - 7\boldsymbol{j} + 11\boldsymbol{k} - 2 \times (\boldsymbol{i} - 2\boldsymbol{j} + 6\boldsymbol{k}) = \boldsymbol{i} - 3\boldsymbol{j} - \boldsymbol{k}$$

$$\boldsymbol{A} \times \boldsymbol{B} = -\boldsymbol{B} \times \boldsymbol{A} = \begin{vmatrix} \boldsymbol{i} & \boldsymbol{j} & \boldsymbol{k} \\ 3 & -7 & 11 \\ 1 & -2 & 6 \end{vmatrix} = -\mathbf{20}\boldsymbol{i} - \mathbf{7}\boldsymbol{j} + \boldsymbol{k}.$$

Summary

In this chapter, we give some preliminary knowledge to learn *Theoretical Mechanics*. The students are required to master such contents as the trigonometric function, concept of distance, scalar, and vector.

Exercises

1.1 $\boldsymbol{a} = (3, 2, -1)$, $\boldsymbol{b} = (1, -1, 2)$, please give the following results:

(1) $\boldsymbol{a} \times \boldsymbol{b}$
(2) $(\boldsymbol{a} + 2\boldsymbol{b}) \times \boldsymbol{b}$
(3) $\boldsymbol{a} \cdot 3\boldsymbol{b}$
(4) $\boldsymbol{b} \times \boldsymbol{a}$
(5) $\boldsymbol{b} \times (3\boldsymbol{a} - 2\boldsymbol{b})$

1.2 $\boldsymbol{a} = (2, -1, 3)$, $\boldsymbol{b} = (1, -1, 3)$, $\boldsymbol{c} = (1, -2, 0)$, please try to present the final results:

(1) $\boldsymbol{a} \times \boldsymbol{b} \cdot \boldsymbol{c}$
(2) $\boldsymbol{a} \cdot \boldsymbol{b}\boldsymbol{c}$
(3) $\boldsymbol{a} \times (\boldsymbol{b} \times \boldsymbol{c})$

(4) $(\boldsymbol{a} \times \boldsymbol{b}) \times \boldsymbol{c}$
(5) $\boldsymbol{c} \times (\boldsymbol{b} \cdot \boldsymbol{c})\boldsymbol{a}$

1.3 $\boldsymbol{A} = (1, -2, 3)$, $\boldsymbol{B} = (3, -2, 4)$, $\boldsymbol{C} = (0, -1, 0)$, and $\boldsymbol{D} = (2, -1, 3)$. Please give the following results:

(1) $\boldsymbol{A} \times \boldsymbol{B} \times \boldsymbol{C} - 2\boldsymbol{D}$
(2) $\boldsymbol{B} \times (\boldsymbol{A} \times \boldsymbol{C}) - \boldsymbol{D} \times \boldsymbol{A}$

Answers

1.1 (1) (3, –7, –5)
(2) (3, –7, –5)
(3) –3
(4) (–3, 7, 5)
(5) (–9, 21, 15)
1.2 (1) 6
(2) (6, –3, 9)
(3) (–8, 20, 12)
(4) (–2, –1, 3)
(5) (–18, –9, 9)
1.3 (1) (0, 2, –4)
(2) (–1, 18, 9)

Part I
Statics

Statics is an ancient terminology from Greece, and the last letter "s" of the word means "science". Therefore, Statics is the science to investigate the balance laws of objects under a force system. It does not deal with motion, acceleration and geometric configuration of objects. The founders of Statics can be traced back to Aristotle, Archimedes, Stevin and L. Poinsot (1777–1859).

Chapter 2
Fundamentals of Statics

Abstract In this chapter, we give some fundamental knowledge on Statics. We will introduce three aspects: concept of force, axioms in Statics, and moment of a force. This is the beginning of Statics and the whole course.

Keywords Force · Axioms in statics · Moment of a force

2.1 Force

Force is the mechanical integration of different objects, which will cause the variation of motion state and deformation of the objects. The first influence is the so-called external effect and the second is the internal effect. However, the main object in this course is the rigid body, which is assumed to never deform throughout this textbook. In other words, our goal is only to examine the equilibrium and motion laws for rigid bodies.

In engineering, we often name force as load, such as wind load, snow load, rain load, seismic or earthquake load, and electromagnetic load. The normally seen forces can be divided into two groups, i.e., the concentrated force and the distributed force. As shown in Fig. 2.1, a concentrated force is modeled as being acted at only one point, and a distributed force is abstracted as being applied on an area or a length span of the object. In the real world, there are no concentrated forces, and "concentrated force" is only a perfect model. When the action area is small enough, such as a needle penetrating into the skin can be considered as a concentrated force. As a consequence, the unit of the concentrated force is N or kN, and that of the distributed force is N/m or kN/m (force per unit length). Notice that $\boldsymbol{P}$ is a vector and $q(x)$ is a scalar in Fig. 2.1, for $q(x)$ denotes the line density of the load.

As is well known, there are three essential elements of the force, namely, magnitude, direction, and point of action. A force $\boldsymbol{F}$ can be expressed as $\boldsymbol{F} = F\boldsymbol{n}$, where F is the magnitude of the force and $\boldsymbol{n}$ is the unit vector. We then have

$$F = |\boldsymbol{F}|.$$

J. Liu, *Lecture Notes on Theoretical Mechanics*,
https://doi.org/10.1007/978-981-13-8035-8_2

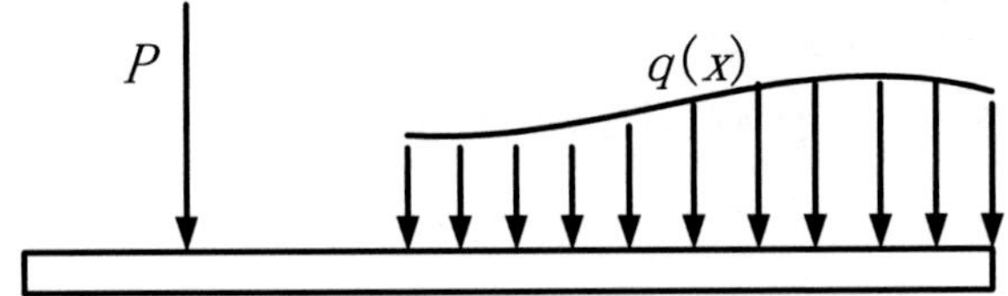

Fig. 2.1 Concentrated force and distributed force

The vector format of the force is not convenient to analyze, and it is normally decomposed in a practical coordinate system. In the Cartesian coordinate system, the force can be written as

$$\boldsymbol{F} = \boldsymbol{X} + \boldsymbol{Y} + \boldsymbol{Z} = X\boldsymbol{i} + Y\boldsymbol{j} + Z\boldsymbol{k},$$

where X, Y, and Z are the three components of the force $\boldsymbol{F}$, which is shown in Fig. 2.2. The three forces are

$$\begin{aligned} \boldsymbol{X} &= X\boldsymbol{i}, \\ \boldsymbol{Y} &= Y\boldsymbol{j}, \\ \boldsymbol{Z} &= Z\boldsymbol{k}. \end{aligned}$$

The angles between the force between x-, y-, and z-axes are α, β, and γ, respectively.

The magnitude of the force is expressed as

$$F = \sqrt{X^2 + Y^2 + Z^2},$$

and

$$\cos\alpha = \frac{X}{F},$$

Fig. 2.2 A force in the Cartesian coordinate system

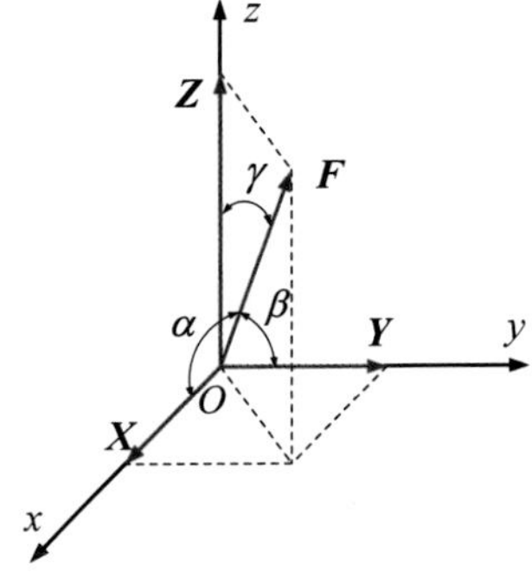

$$\cos\beta = \frac{Y}{F},$$
$$\cos\gamma = \frac{Z}{F}.$$

The force composition must satisfy the parallelogram law. For example, if $\boldsymbol{R}$ is the resultant force of two forces $\boldsymbol{F}_1$ and $\boldsymbol{F}_2$, as shown in Fig. 2.3, the diagonal line of the parallelogram denotes the resultant force in geometry. The equivalent method is the vector triangle law, which is schematized in Fig. 2.4.

The relation between the resultant force and the components is

$$\boldsymbol{R} = \boldsymbol{F}_1 + \boldsymbol{F}_2,$$

where

$$\begin{aligned}\boldsymbol{F}_1 &= X_1\boldsymbol{i} + Y_1\boldsymbol{j} + Z_1\boldsymbol{k},\\ \boldsymbol{F}_2 &= X_2\boldsymbol{i} + Y_2\boldsymbol{j} + Z_2\boldsymbol{k},\\ \boldsymbol{R} &= R_x\boldsymbol{i} + R_y\boldsymbol{j} + R_z\boldsymbol{k}.\end{aligned}$$

According to the operation rule of the vector, one has

$$\begin{aligned}R_x &= X_1 + X_2,\\ R_y &= Y_1 + Y_2,\\ R_z &= Z_1 + Z_2.\end{aligned}$$

In essence, this relation means that the components among the resultant force and the two forces are corresponding to each other. The magnitude of the resultant force is derived as

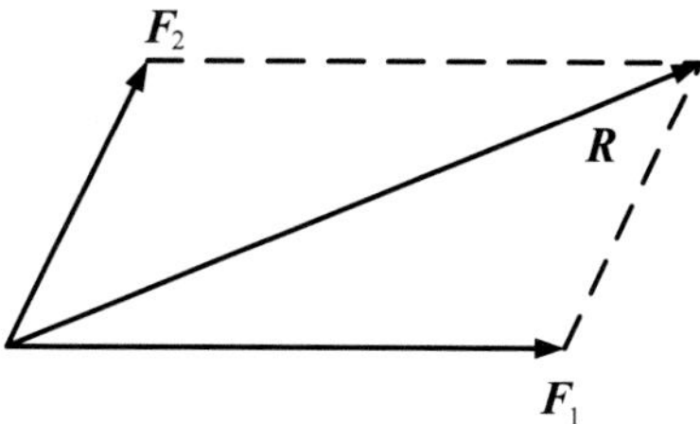

Fig. 2.3 Parallelogram law

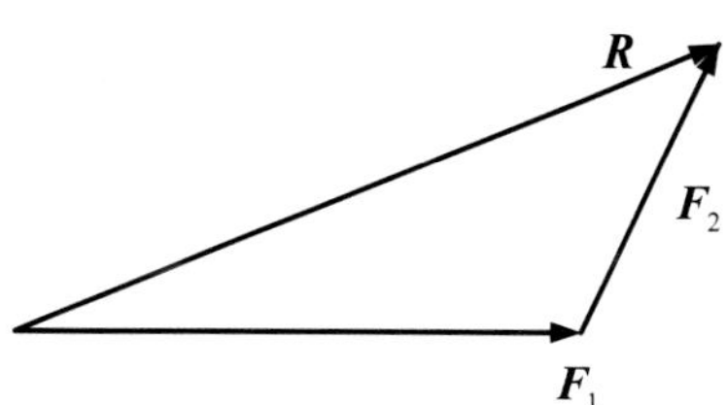

Fig. 2.4 Vector triangle law

$$R = \sqrt{R_x^2 + R_y^2 + R_z^2},$$

and

$$\cos\alpha = \frac{R_x}{R},$$
$$\cos\beta = \frac{R_y}{R},$$
$$\cos\gamma = \frac{R_z}{R}.$$

2.2 Axioms in Statics

Based upon a lot of observations and practices, scientists and engineers have concluded several axioms in Statics. They have already been proved to be true in practice, and there is no need to prove them from the mathematical viewpoint. We only master the conclusions and don't deal with the details.

Axiom 1 Two forces in equilibrium axiom
If two forces acting on a rigid body are in equilibrium, they must have the same magnitude, opposite directions, and they are along the same action of line. Otherwise, the rigid body cannot be balanced. In this case, the rigid body can be named as a two-force member or two-force bar, as shown in Fig. 2.5. The key point is that there are **only** two points on the rigid body enduring forces, and the final resultant forces at these two points have the above relations. We always remember that two points determine one line, and the two forces must be collinear.

The relation of the two forces in Fig. 2.5 is

$$\boldsymbol{F} = -\boldsymbol{F}',$$
$$F = F'.$$

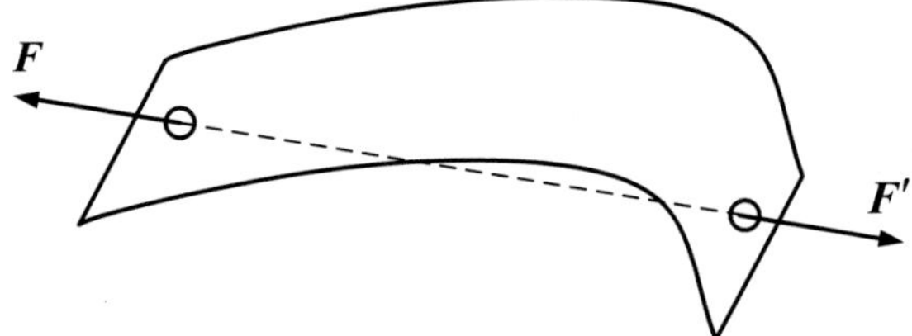

Fig. 2.5 Two-force bar

Axiom 2 Axiom of addition or subtraction of a force group in equilibrium
If an equilibrium force group is applied or removed from a rigid body, the final effect on the rigid body does not change as before. For example, as shown in Fig. 2.6a, an object lying on the desk is in equilibrium. If we apply another force group in equilibrium on the object, where the two forces are collinear, with opposite directions and of the same magnitude, as shown in Fig. 2.6b, the original rigid body must be balanced. On the contrary, if the force group in equilibrium is removed, the object does not change.

From this axiom, we can also prove that a force acting on a rigid body can transmit along its action line without changing the state of the body, as shown in Fig. 2.7. Initially, the action point of the force is at point A, and it can be moved along the action line of the force, directly to point B. In this process, the state of the rigid body does not alter.

Axiom 3 Axiom of three concurrent forces in equilibrium
If a rigid body is in equilibrium under the action of three forces, and two of them are crossed at one point, then the third force must pass through this point, as schematized in Fig. 2.8. Two of the forces ($\boldsymbol{F}_1$ and $\boldsymbol{F}_2$) can be further composed to one force, and the new resultant force ($\boldsymbol{F}_1 + \boldsymbol{F}_2$) and the third force $\boldsymbol{F}_3$ constitute in a new force system in equilibrium. Therefore, according to Axiom 1, they should be collinear, and then the current axiom can be proved.

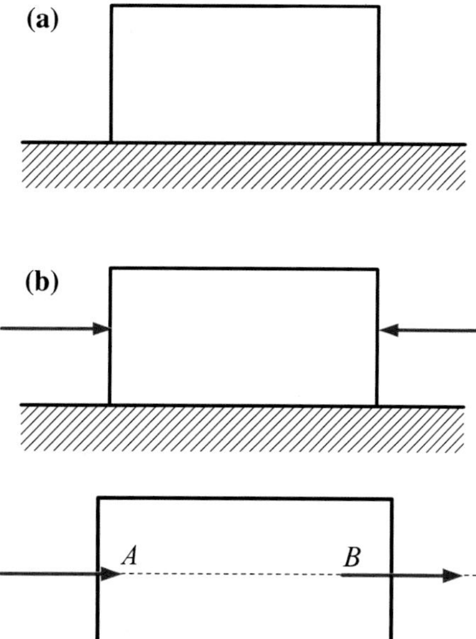

Fig. 2.6 Addition of a force group in equilibrium

Fig. 2.7 A force transmits along a line

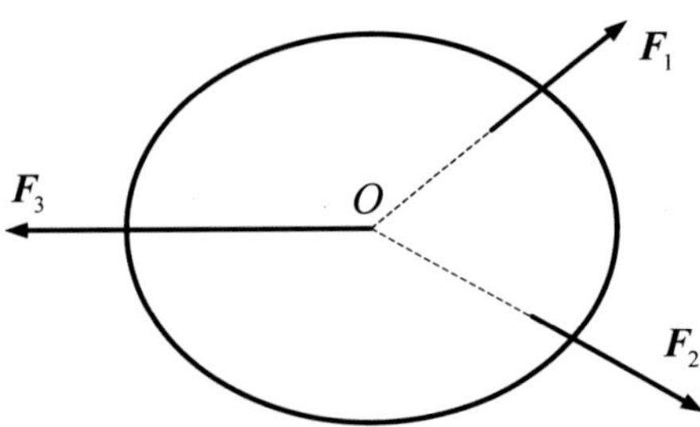

Fig. 2.8 Three concurrent forces in equilibrium

2.3 Moment of a Force

From the middle school stage, we have already known the concept of moment or torque, which is defined as "force times distance". During this situation, the model is normally in planar case, and the support point is actually equivalent to an axis normal to the plane. Herein, we give the more general definition of the moment of force with respect to an axis, and it depicts the rotation effect of a force to an axis.

The vector $\boldsymbol{r}$ is built from the origin O, which is named as the displacement vector or radius vector. In the Cartesian coordinate system, it can be expressed as

$$\boldsymbol{r} = x\boldsymbol{i} + y\boldsymbol{j} + z\boldsymbol{k},$$

where (x,y,z) is the coordinate of point M. The force $\boldsymbol{F}$ is applied at point M, which has three components shown in Fig. 2.9. It is easily seen that, if a force passes through an axis or it is parallel to this axis, then it has no moment with respect to the axis. From the geometric point of view, the component $\boldsymbol{Z}$ is parallel to the axis z, and then it has no contribution to the rotation of z-axis. Therefore, following the concept of "force times distance", we define the moment of force to z-axis

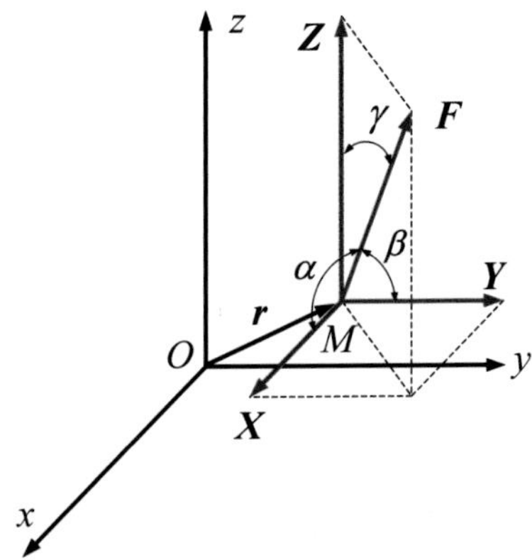

Fig. 2.9 Moment of force to axes

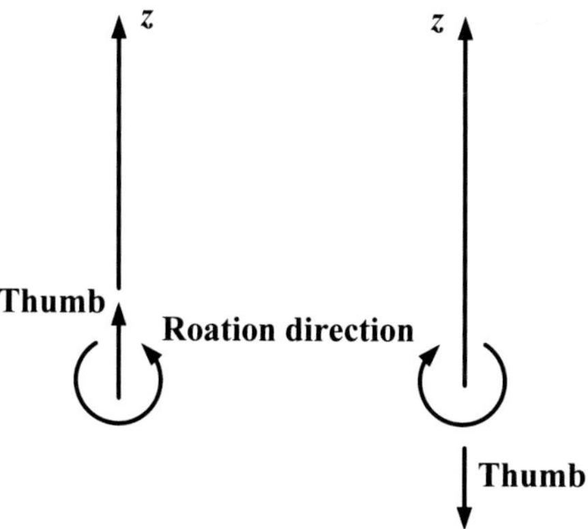

Fig. 2.10 Moment direction

$$M_z(\boldsymbol{F}) = xY - yX.$$

The minus sign in the above formula should be further stressed. From the figure, one can see that the distance between $\boldsymbol{X}$ and z-axis is y, and that between $\boldsymbol{Y}$ and z-axis is x. We here take the right-hand screw rule, indicating that if the thumb direction of the moment is the same as that of z-axis, the moment is positive, and otherwise it is negative. From Fig. 2.9, one can see that $\boldsymbol{Y}$ causes the positive moment with the distance x, and $\boldsymbol{X}$ causes the negative moment with the distance y. To clearly express the rotation effect of the moment, we can refer Fig. 2.10. The curved arrowhead represents the rotation or rolling direction of the moment, and then one can judge the thumb direction according to the right-hand screw rule. If the thumb direction is the same as that of z-axis, it is positive; otherwise, it is negative.

Following the above definitions, one can get the other moments to the axes

$$M_x(\boldsymbol{F}) = yZ - zY,$$
$$M_y(\boldsymbol{F}) = xY - yX.$$

In the planar case, such as in the o-xy coordinate system, the moment of a force to axis z degenerates to the moment to point o. In this situation, we define that the anticlockwise rotation of the moment is positive, and the clockwise direction is negative. In fact, this conclusion coincides with the right-handed rule, where the point o can be imagined as the projection of the normal line of the plane, which is the axis. In this case, the moment of the force with respect point o is

$$M_o(\boldsymbol{F}) = M_z(\boldsymbol{F}) = xY - yX.$$

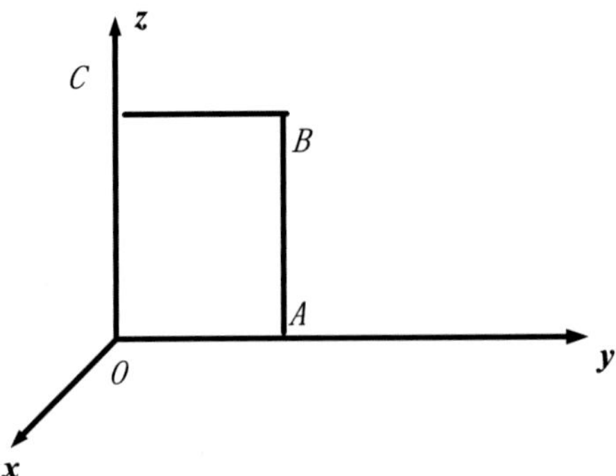

Fig. 2.11 Rotation of the door

Example 1 As shown in Fig. 2.11, $OABC$ is a rectangular door in o-yz plane, where a force is applied at point B to close the door. $BC = 3$ m, $AB = 4$ m, $\boldsymbol{F} = -\boldsymbol{i} + 4\boldsymbol{k}$ (unit: N). Please give the moment to the z-axis $M_z(\boldsymbol{F})$.

Answer: The coordinate of point B is (0, 3, 4), $X = -1$, $Y = 0$, $Z = 4$.

That is to say, the displacement vector of point B is

$$\boldsymbol{r} = 3\boldsymbol{j} + 4\boldsymbol{k}$$

The force can be expressed as

$$\boldsymbol{F} = -\boldsymbol{i} + 4\boldsymbol{k}$$

The moment of force to z-axis is

$$M_z(\boldsymbol{F}) = xY - yX = 3\,\mathrm{N}\cdot\mathrm{m}$$

We can also decompose the force into the three axes. It has two components, but the one in the z-direction has no contribution to $M_z(\boldsymbol{F})$. Consequently, there is only one contribution from the component in the y-direction. According to "force times distance", the moment should be $3\,\mathrm{N}\cdot\mathrm{m}$.

Example 2 In the planar case, please give the moment of force to point o (Fig. 2.12).

Answer: In the two-dimensional case, $M_o(\boldsymbol{F}) = M_z(\boldsymbol{F}) = xY - yX$. The sign of the moment can be verified: if it is anticlockwise, then the sign is positive; otherwise, it is negative.

Another important concept is the moment of force with respect to a point. It introduces a new vector besides the displacement vector and force vector, which is given as

$$\boldsymbol{M}_o(\boldsymbol{F}) = \boldsymbol{r} \times \boldsymbol{F}.$$

According to the definition of cross product, the direction of the new moment is normal to the plane determined by the force and displacement vector, as shown in Fig. 2.13. In essence, this is the real application of the right-handed rule.

The expansion of the moment of force to a point is

$$\boldsymbol{M}_o(\boldsymbol{F}) = \begin{vmatrix} \boldsymbol{i} & \boldsymbol{j} & \boldsymbol{k} \\ x & y & z \\ X & Y & Z \end{vmatrix} = (xY - yX)\boldsymbol{k} + (yZ - zY)\boldsymbol{i} + (zX - xZ)\boldsymbol{j}.$$

Clearly, one has

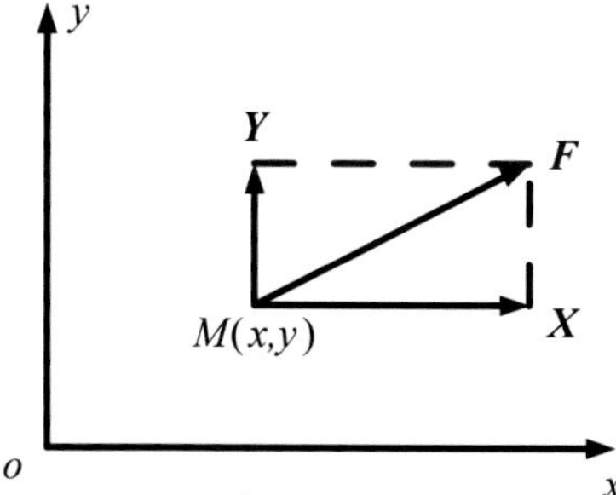

Fig. 2.12 Moment in a plane

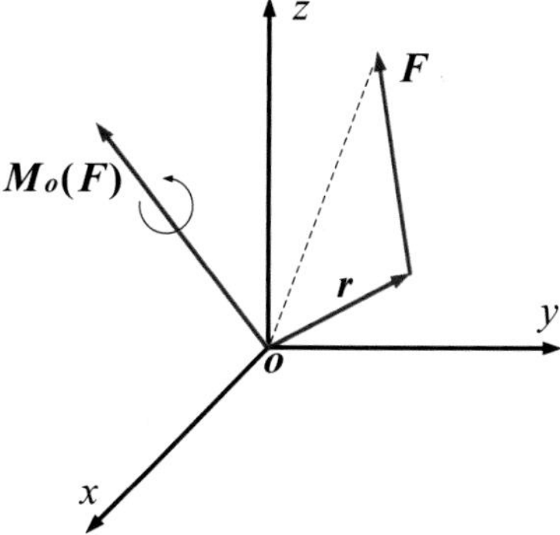

Fig. 2.13 Moment of force to a point

$$\boldsymbol{M}_o(\boldsymbol{F}) = M_x(\boldsymbol{F})\boldsymbol{i} + M_y(\boldsymbol{F})\boldsymbol{j} + M_z(\boldsymbol{F})\boldsymbol{k},$$

meaning that the three moments of the force to the axes all have contributions to the moment of a force to the point. In fact, this relation provides a novel method to solve the moment of a force to one point. Maybe it is intractable for most of us to calculate the cross product due to complex operations, and then we can first calculate the moment of a force to the three axes to get the final result.

For the problem in Example 1 of this chapter, one has

$$\boldsymbol{M}_o(\boldsymbol{F}) = \begin{vmatrix} \boldsymbol{i} & \boldsymbol{j} & \boldsymbol{k} \\ x & y & z \\ X & Y & Z \end{vmatrix} = \begin{vmatrix} \boldsymbol{i} & \boldsymbol{j} & \boldsymbol{k} \\ 0 & 3 & 4 \\ -1 & 0 & 4 \end{vmatrix} = 12\boldsymbol{i} + 4\boldsymbol{j} + 3\boldsymbol{k}.$$

Then we have

$$M_x(\boldsymbol{F}) = 12\,\mathrm{N}\cdot\mathrm{m},\ M_y(\boldsymbol{F}) = 4\,\mathrm{N}\cdot\mathrm{m},\ M_z(\boldsymbol{F}) = 3\,\mathrm{N}\cdot\mathrm{m}.$$

For the problem in Example 2 of this chapter, one has

$$\boldsymbol{M}_o(\boldsymbol{F}) = \begin{vmatrix} \boldsymbol{i} & \boldsymbol{j} & \boldsymbol{k} \\ x & y & 0 \\ X & Y & 0 \end{vmatrix} = (xY - yX)\boldsymbol{k},$$

and there is only one component $M_z(\boldsymbol{F}) = xY - yX$.

Summary

In this chapter, some fundamental knowledge on Statics is introduced. The content includes three aspects: concept of force, axioms in Statics, and moment of a force. This is the fundamental knowledge of Statics and the whole course.

Exercises

2.1 The figure is shown in Fig. 2.14. Determine the moment of the force $\boldsymbol{F}$

(1) about point C
(2) about axis x

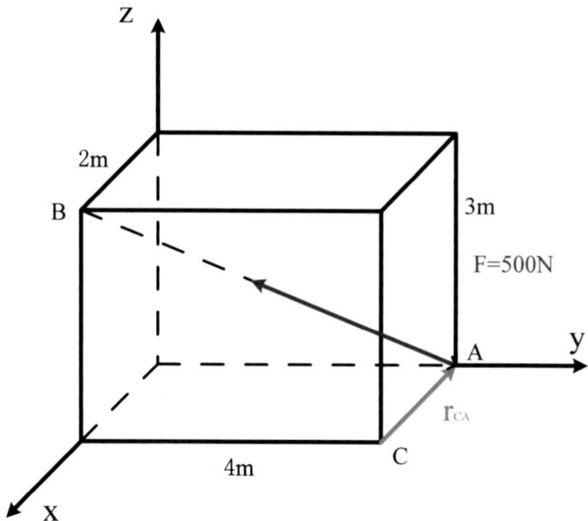

Fig. 2.14 Moment of a force

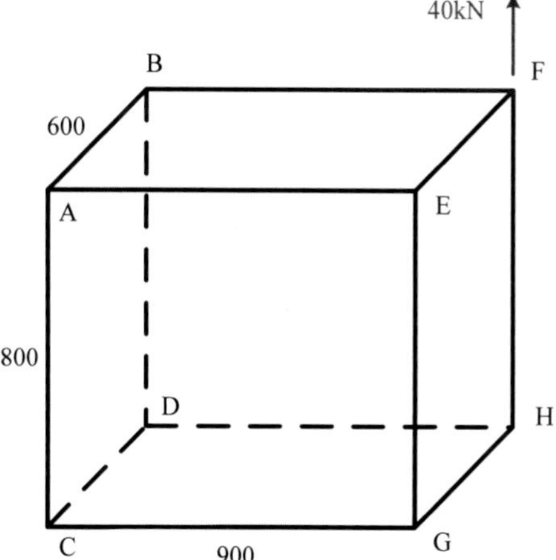

Fig. 2.15 Moment of a force

2.2 Determine the moments of the force about the following axes (as shown in Fig. 2.15):

(1) *AB* (2) *CD* (3) *CG* (4) *CH* (5) *EG*

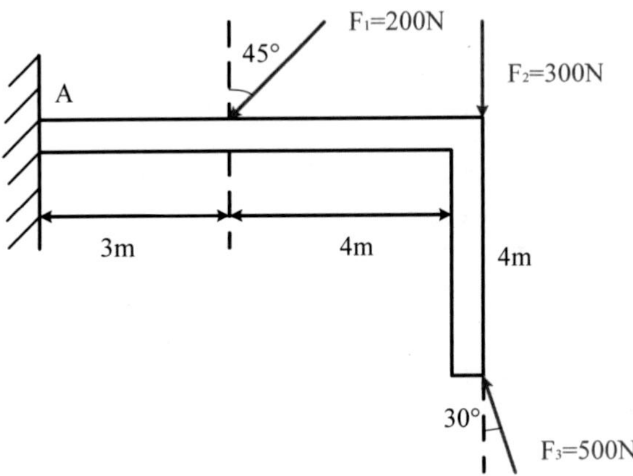

Fig. 2.16 Planar moment of a force

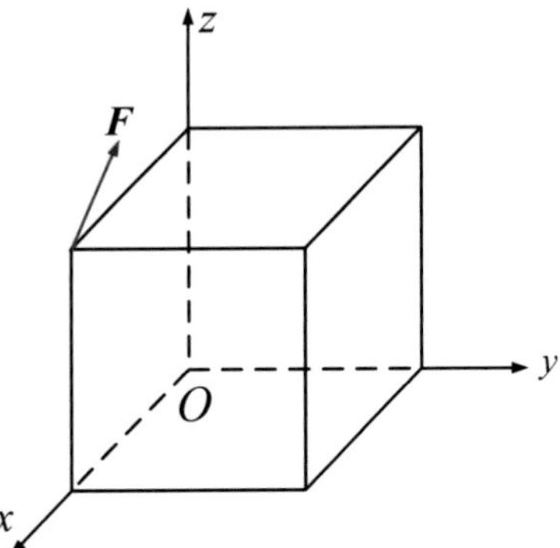

Fig. 2.17 A cuboid with one force

2.3 Determine the total moments of all forces about point A (as shown in Fig. 2.16).

2.4 As shown in the Cartesian coordinate system in Fig. 2.17, there is a cuboid with the side length 1 cm. There is a force $\boldsymbol{F} = -4\boldsymbol{i} + 5\boldsymbol{k}$ (unit: N) applied at one corner. Please give the expressions of the moment of the force to z-axis, and the moment of the force to point O.

Answers

2.1 (1) $557.09\boldsymbol{j} + 742.78\boldsymbol{k}\,\mathrm{N \cdot m}$ (2) $1114.17\boldsymbol{i}\,\mathrm{N \cdot m}$
2.2 (1) $36\,\mathrm{kN \cdot m}$ (2) $36\,\mathrm{kN \cdot m}$ (3) $24\,\mathrm{kN \cdot m}$ (4) $0\,\mathrm{kN \cdot m}$ (5) $0\,\mathrm{kN\,m}$
2.3 $M_A = 493.18\,\mathrm{kN\,m}$
2.4 $M_z = 0,\ M_O = -0.09\boldsymbol{j}\,\mathrm{N \cdot m}.$

Chapter 3
Force Analysis

Abstract In this chapter, we will learn how to perform the force analysis. First, the concept of constraint and reactive force is introduced, and then the steps on force analysis are given.

Keywords Constraint · Reactive force · Force analysis · Free body diagram

3.1 Constraint

A rigid body generally is subjected to two types of forces, i.e., one is the active force, such as gravity, wind load, snow load, rain load, and electromagnetic force, and the other is the reactive force from the constraints on the rigid body. For example, a bird flying in the sky only undergoes the gravity if the air resistance force is neglected. In this case, it has no constraints and can be named as a free body. A constraint is to resist the movement of the rigid body, which therefore can induce forces to hold back the object. In engineering applications, there are some normally seen constraints, which produce constraint forces or reactive forces. The classical constraints are formulated as follows.

(1) **Flexible cable**

An object is often constrained by a flexible cable, which means that the cable provides a reactive force to confine the object. As shown in Fig. 3.1, a ball is hung by a rope on the ceiling. If there is no force from the rope, the ball will go downward to the ground. This fact manifests that a reactive force from the cable must exist to balance the gravity of the ball. We first relieve the constraint, only investigating the object. This process is called isolation, and the object is called a free body now. The force from the rope deviates from the object, along the contraction direction of the rope. We use the symbol $\boldsymbol{T}$ to represent the force, with the meaning of tension.

Similarly, the belt of a wheel can transport some objects from one position to another, and the belt is normally in tension. We analyze the wheel, and it endures the pulling force from the belt, as shown in Fig. 3.2. The tension also deviates from the wheel, along the contraction direction of the belt.

J. Liu, *Lecture Notes on Theoretical Mechanics*,
https://doi.org/10.1007/978-981-13-8035-8_3

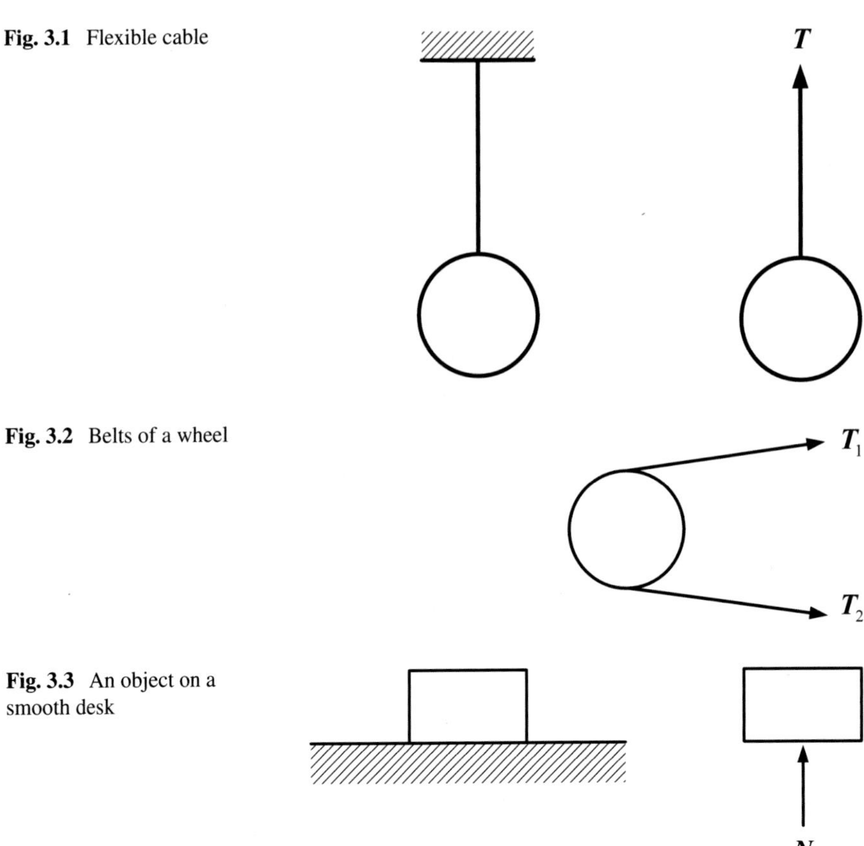

Fig. 3.1 Flexible cable

Fig. 3.2 Belts of a wheel

Fig. 3.3 An object on a smooth desk

(2) **Smooth contact surface**

As shown in Fig. 3.3, an object is lying on a desk. If there is no supporting force from the desk, the object will go downward to the ground. This means the desk provides a reactive force to ensure the constraint of the object. This kind of constraint is called smooth contact surface constraint, where the friction is neglected. The direction of the reactive force from the contact surface is normal to this surface, pointing inside to the object. The action line of the reactive force can also be determined considering that it is perpendicular to the tangential line of the contact surface. As shown in Fig. 3.4, an object is put on a curved surface. We first relieve the constraint, leaving only a free body. To determine the direction of the reactive force from the curved surface, we can draw the tangential line between the ball and the curved surface. Then the force is vertical to this tangential line.

Figure 3.5 shows the contact of two gears, where the teeth are contacting with each other to transmit torques. Select one tooth, and it undergoes the force from the neighborhood tooth. Relieving the constraint of the neighborhood tooth, the reactive

Fig. 3.4 An object on a curved surface

Fig. 3.5 Gears

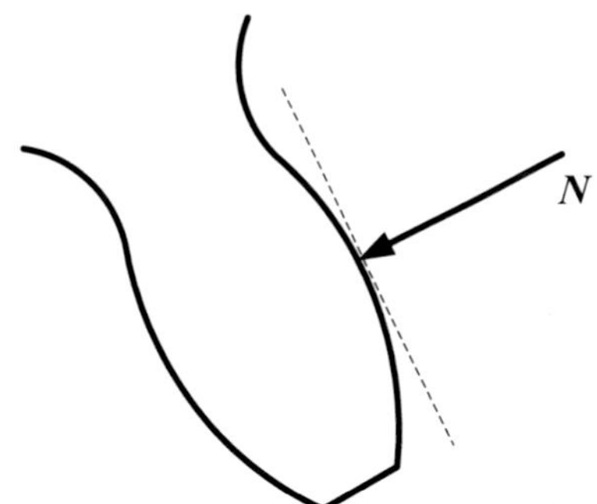

Fig. 3.6 Contact force of a tooth

force of the first tooth is perpendicular to the tangent line of the contact surface, as demonstrated in Fig. 3.6.

Another example is a slender bar put in a trough. There are three contact points which can produce reactive forces. These reactive forces are from the smooth contact surfaces, which can be dictated as the schematic displayed in Fig. 3.7. At point A, a point from the bar is contacting with a planar surface, and the tangential line is of the

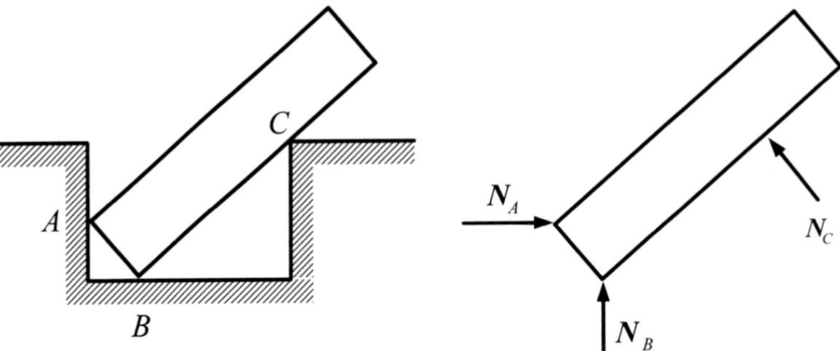

Fig. 3.7 A bar in a trough

same direction with the surface. Point B is similar to A. At point C, a point from the trough is contacting with the bar surface, and the tangential line can also be marked.

(3) **Hinged or pinned support**

If a rod is fixed by the ground, and it can also slightly rotate at this point, this kind of constraint is called hinged or pinned support, as shown in Fig. 3.8. In fact, there are two holes on the ground and the bar, respectively, which are matched by a cylindrical pin. However, in the process of analysis, the pin can be thought to be fixed on the ground or on the bar, and we only consider the match of the two objects, i.e., the ground (or the other bar) and the selected bar. As the bar can rotate with respect to the hinged point O, it will endure the reactive force from the ground. It is clear that this is a problem of contact surface, where the reactive force is normal to the tangential line of the contact surface. However, the bar can rotate at any angle at the hinged point, and the direction of the reactive force can't be a determined one. Therefore, we decompose the reactive force in two directions, i.e., $\boldsymbol{X}_O$ and $\boldsymbol{Y}_O$, respectively.

The similar case can happen when two bars are hinged together by a cylindrical pin, as schematized in Fig. 3.9. Each bar can rotate with respect to the hinged point, and therefore the direction of the reactive for each bar is not a determined one, then it can be expressed by the two components along the x- and y-coordinate axes, respectively.

Fig. 3.8 A hinged bar on the ground

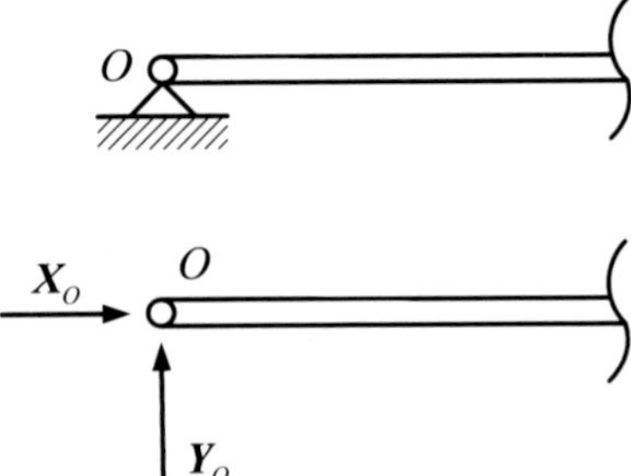

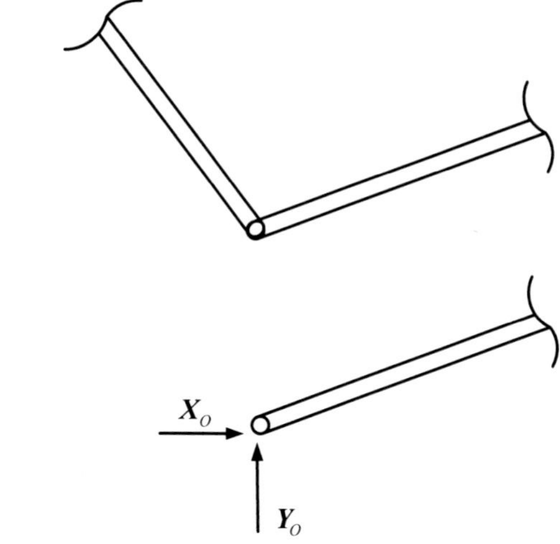

Fig. 3.9 A hinged bar

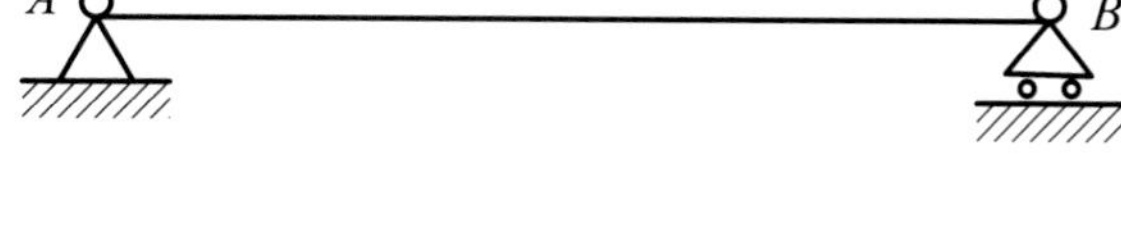

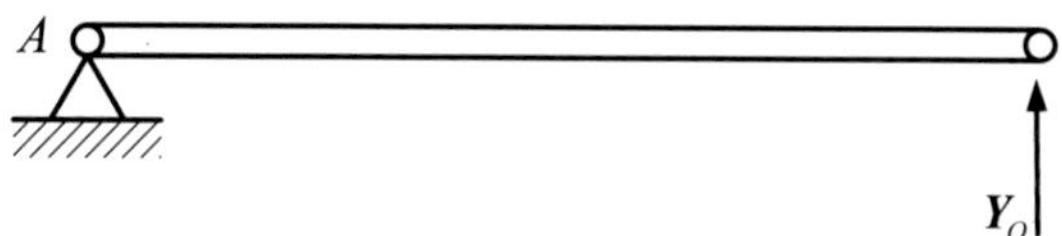

Fig. 3.10 Roller support

(4) Roller support

Another special constraint is the roller support, which is shown at point B in Fig. 3.10. The constraint is similar to a small car, which can roll in the horizontal direction, but it can't move in the vertical direction. As a result, the constraint can only provide one component reactive force Y_O, which is shown in Fig. 3.10. This structure can be also expressed in two other formats, which are shown in Fig. 3.11. The roller support is especially important in engineering. For example, if we design a bridge, which can be modeled as a beam with two hinged ends, then one end should be designed as a roller support. If both ends are fixed as hinged constraints, then the temperature can induce elongation or compression, and the bridge can't match this deformation. The roller support can make the bridge slightly move in the horizontal direction, and can overcome the influence of temperature.

(5) Smooth ball and socket joint

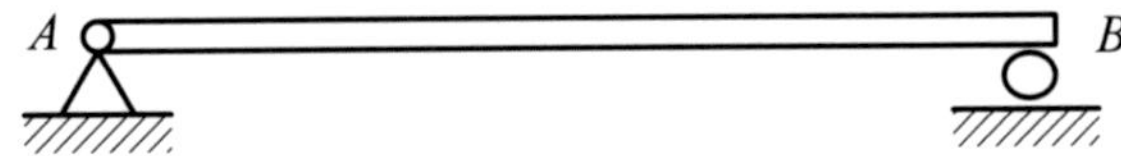

Fig. 3.11 Roller support

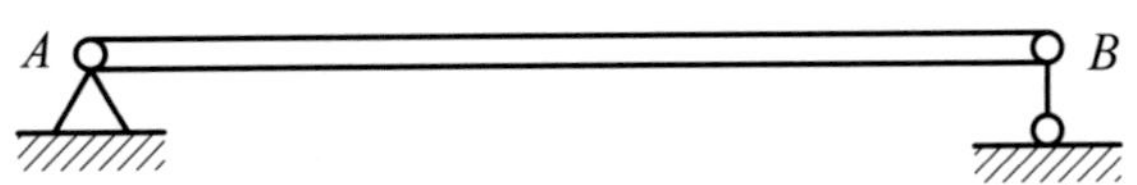

The cylindrical hinge can be viewed as a two-dimensional constraint, and the smooth ball and socket joint can be thought of as a three-dimensional constraint, as schematized in Fig. 3.12. The small ball can rotate in the socket freely, where the reactive force can be produced in any direction. In the Cartesian coordinate system, the reactive force can be expressed by the three components in x-, y- and z-directions. We therefore use the three components to express the reactive force from the socket.

(6) **Bearing**

A bearing can support a shaft, which rotates in it. Two types of bearings are normally seen in industry, i.e., the centripetal bearing and thrust bearing, which are schematized in Figs. 3.13 and 3.14, respectively. The centripetal bearing can't resist the movement of the shaft in the y-direction, but the thrust bearing can. As a consequence, the reactive forces from the two types of bearings are, respectively, displayed in Figs. 3.13 and 3.14.

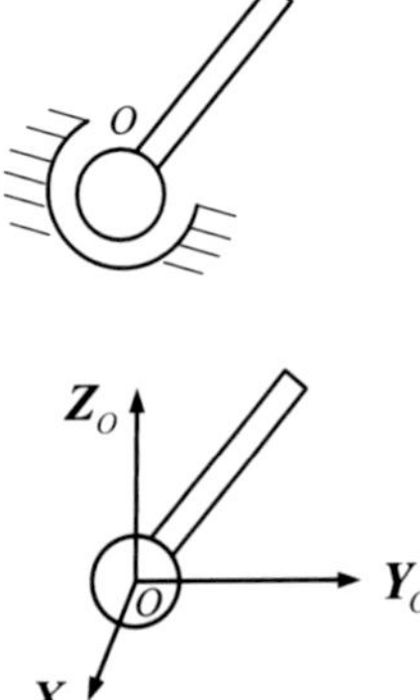

Fig. 3.12 Smooth ball and socket joint

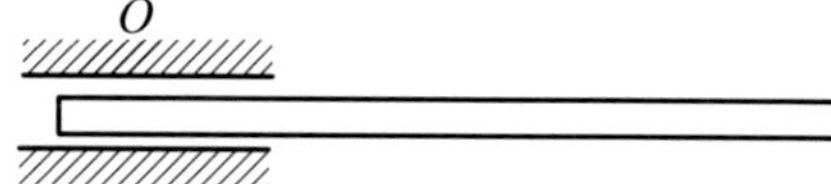

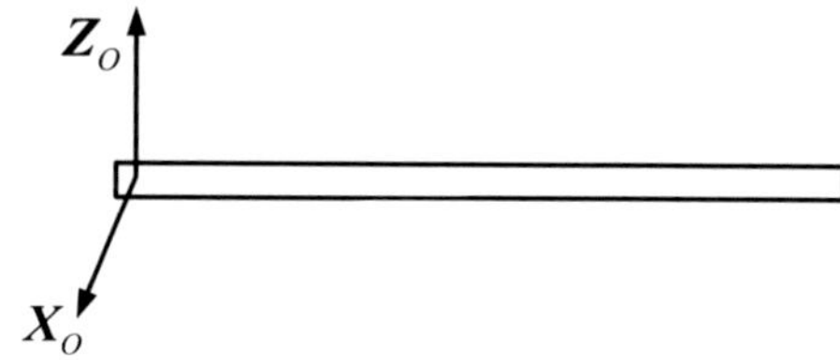

Fig. 3.13 Centripetal bearing

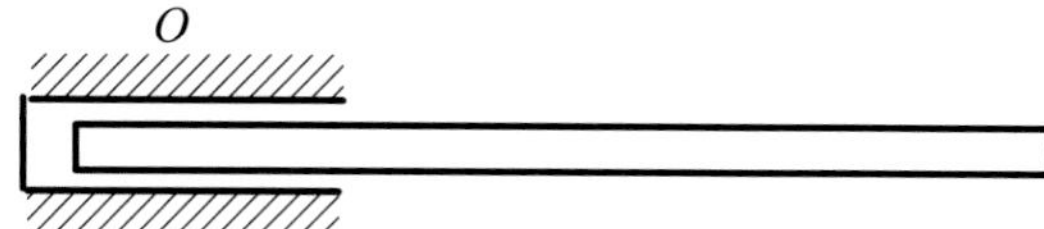

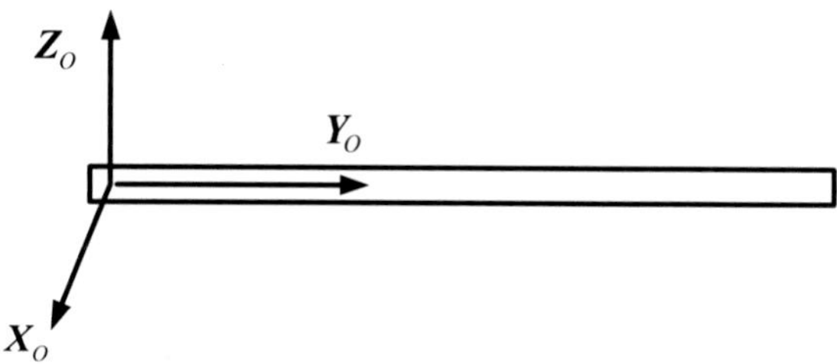

Fig. 3.14 Thrust bearing

3.2 Force Analysis

One of the most important aspects in *Theoretical Mechanics* is how to carry out the force analysis. This is the first and especially important step to solve a static and dynamic problem. We should remember the following steps:

(1) Select the object from the system and isolate it from the others. There are normally a lot of objects in a system, and you should select the most interested one to analyze. After the object selection, you need to separate it as an isolated body, which is termed as "free body". If there are two-force bars, we should first analyze it, as the direction of the force can be determined.
(2) Analyze all the forces, including the active and reactive forces acting on the object.

(3) Draw the free body diagram. Add all the active and reactive forces at the proper positions.

We take the first example to analyze, which is shown in Fig. 3.15. A disk is clamped by the left wall and an inclined bar, and the bar is tensiled by a rope which is fixed on the wall at the other end. We first select the disk as the object, and there are three forces acting on it, i.e., the gravity and two reactive forces from the smooth contact surfaces at point E and D. According to the definition of the smooth contact surface constraint, the reactive force $\boldsymbol{N}_{\mathrm{E}}$ is normal to the wall surface and the reactive force $\boldsymbol{N}_{\mathrm{D}}$ is normal to the bar. According to the basic axiom in Statics, the three forces should be crossed at one point O, as shown in Fig. 3.16.

We then consider the slender bar AB, which endures reactive forces at three points. At point A, the constraint is the cylindrical hinge, which provides two reactive forces in the x- and y-directions, respectively. At point D, the reactive force is in the opposite direction of the force $\boldsymbol{N}_{\mathrm{D}}$, which obeys Newton's third law, so we normally use the symbol $\boldsymbol{N}'_{\mathrm{D}}$ to represent the reaction. At point B, the constraint is the flexible constraint, which is depicted in Fig. 3.15. However, at point A, the two component forces can be combined into one resultant force, which satisfies the condition that the three forces at point A, B, and D cross at one point, as drawn in Fig. 3.17.

Another example is shown in Fig. 3.18, where a slender bar is located at a wall corner, which is tied by a flexible rope at point B. We follow the above force analysis steps, and the free body diagram is displayed in Fig. 3.19.

The third example is shown in Fig. 3.20, where two bars are hinged at point C, experiencing two active forces, including one distributed force and one concentrated force.

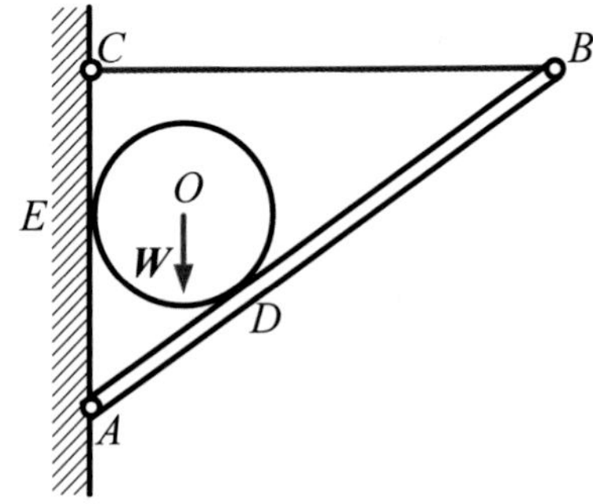

Fig. 3.15 A rigid body system in equilibrium

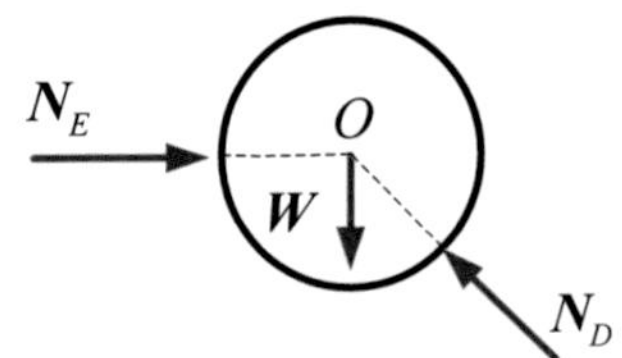

Fig. 3.16 Force analysis

Fig. 3.17 Force analysis

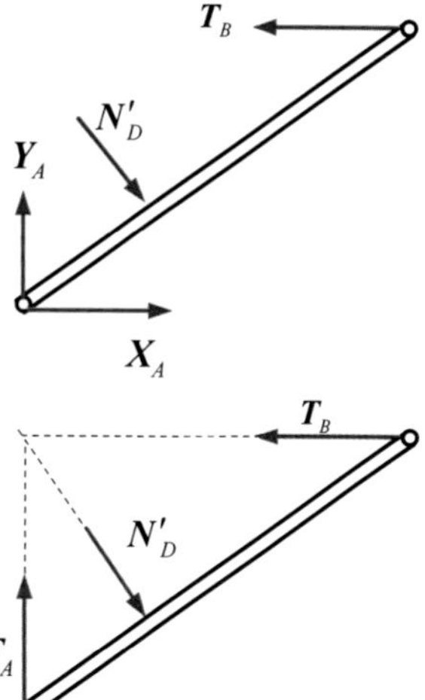

Fig. 3.18 A bar in equilibrium

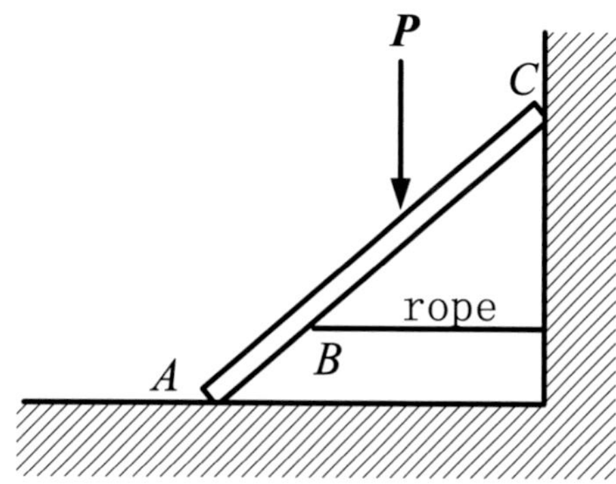

Fig. 3.19 Force analysis

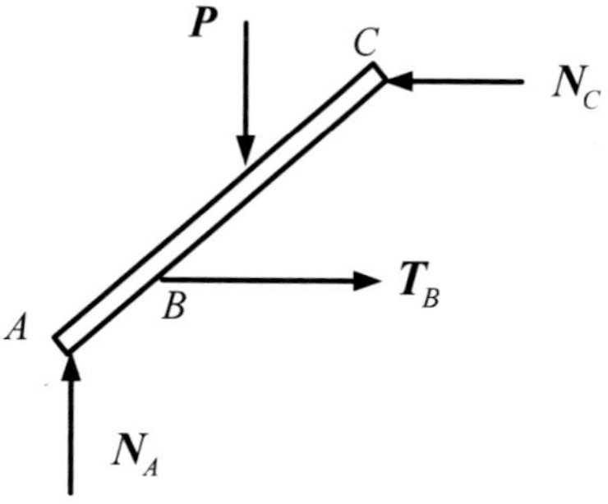

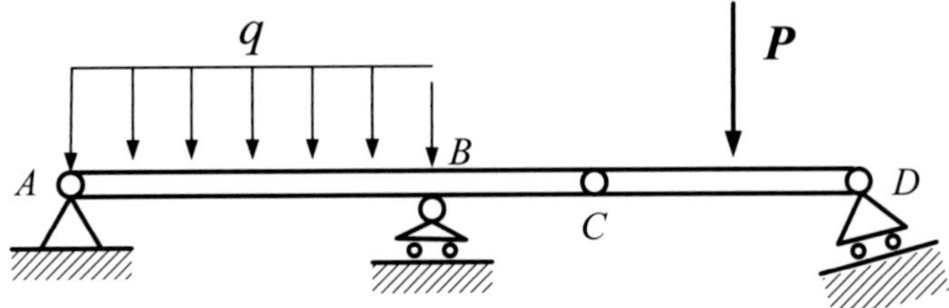

Fig. 3.20 Two bars hinged together

Fig. 3.21 Force analysis

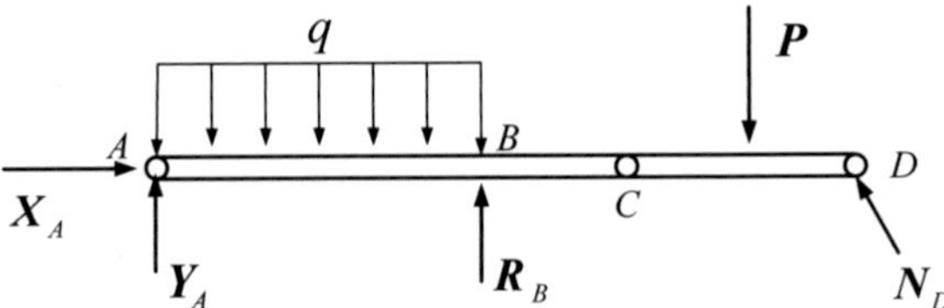

Obviously, there are two rigid bodies in this system, and we should analyze them one by one. We can select the objects as the total system, *AC* bar and *CD* bar. First, we can select the total system as the object. In fact, how to select the object in practice depends on how to solve the problem most efficiently. The free body diagram of the total system is shown in Fig. 3.21.

We first add the two active forces on the free body of the total system. Besides, at point *A*, the constraint type is the cylindrical hinge, which produces two reactive forces. At point *B*, the roller support only provides one reactive force, and at point *D*, the smooth contact surface provides one reactive force normal to the surface. Although point *C* is the cylindrical hinge, its reactive forces are from *AB* bar and *CD* bar, and they are actually the action and reaction, whose resultant force is zero when we consider the full system.

For instance, we can separate the two bars to clearly see the reactive forces at point *C*, as shown in Fig. 3.22. In this case, the bar *CD* can provide two reactive forces acting at point *C* on the bar *AC*. Correspondingly, the bar *AC* can produce reaction to bar *CD*, and it can provide two reactive forces $\boldsymbol{X}'_C$ and $\boldsymbol{Y}'_C$, which are in the opposite directions of the forces $\boldsymbol{X}_C$ and $\boldsymbol{Y}_C$. It should be mentioned that we later always use the symbol prime to represent the reaction force.

Summary

In this chapter, the task is to learn how to perform the force analysis. The concept of constraint and reactive force, and the steps on force analysis must be mastered.

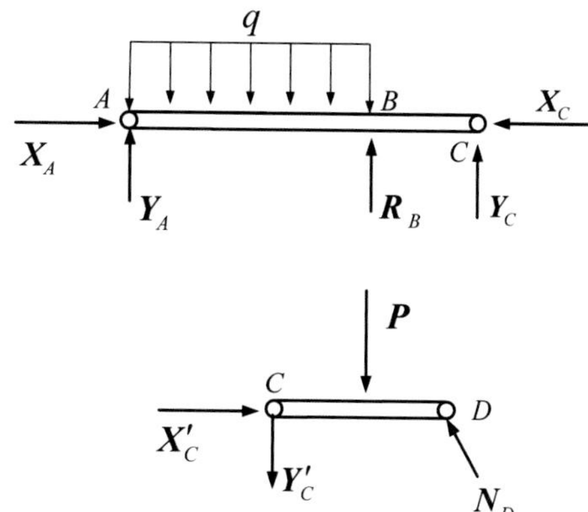

Fig. 3.22 Force analysis

Exercises

3.1 Perform the force analysis about the slender bar *AB* without gravity, and the analysis on the disk with mass *m*, as shown in Fig. 3.23. The bar is fixed by a rope at point *C*, and the disk contacts with the bar at points *D* and *E*. The contact surfaces are both smooth.

3.2 In Fig. 3.24, please perform the force analysis about the bar *AB*, which contacts with a ball at point *C*. The contact surfaces and points are all smooth.

3.3 A cantilever is clamped at point *A*, where a couple *m* and a uniformly distributed force *q* are applied on the beam, as shown in Fig. 3.25. Please perform the force analysis on the beam.

3.4 As shown in Fig. 3.26, there is one bar *AD* hinged with one disk without weight. Point *B* and *C* are both hinged points. A concentrated force is applied at point *D*. Please make force analysis on the bar and the disk.

3.5 As shown in Fig. 3.27, a bar is lying on a wall corner, linked by one rope *BD*. A concentrated force ***F*** is applied at point *E* on the bar. The contact surfaces are smooth. Please make the force analysis on the bar.

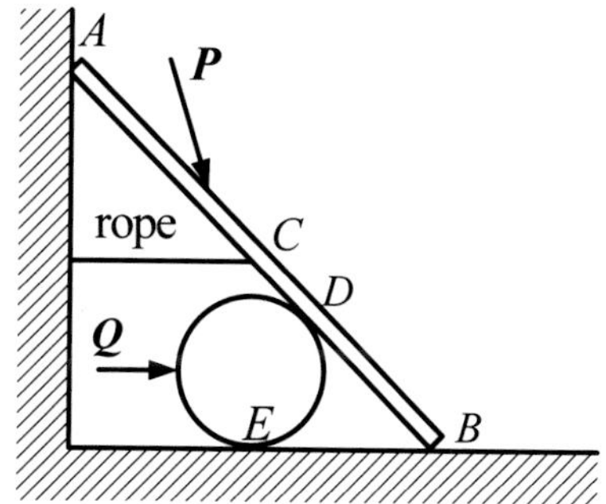

Fig. 3.23 One bar and one disk

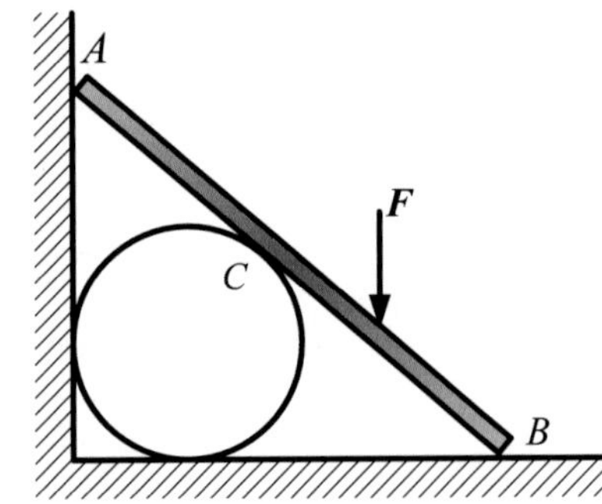

Fig. 3.24 One bar contacting with one disk

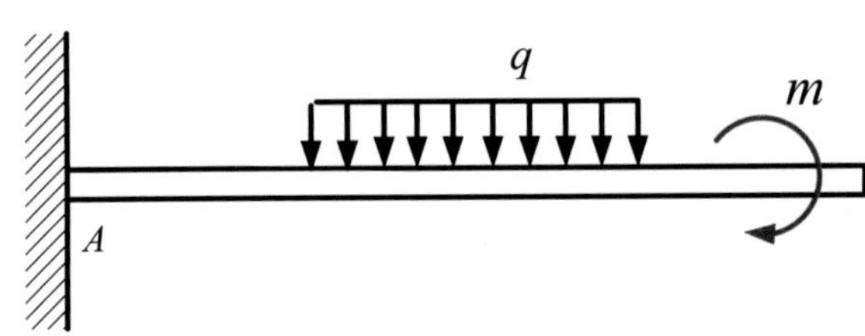

Fig. 3.25 One cantilever

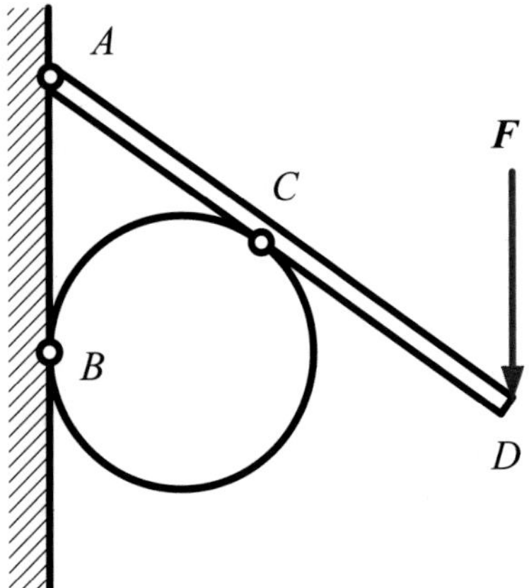

Fig. 3.26 One bar hinged with one disk without weight

3.6 As shown in Fig. 3.28, a system includes three bars, i.e., *AB*, *CD*, and *BE*. A horizontal force $\boldsymbol{F}$ is applied at point *A*. Please perform the force analyses for all of these bars.

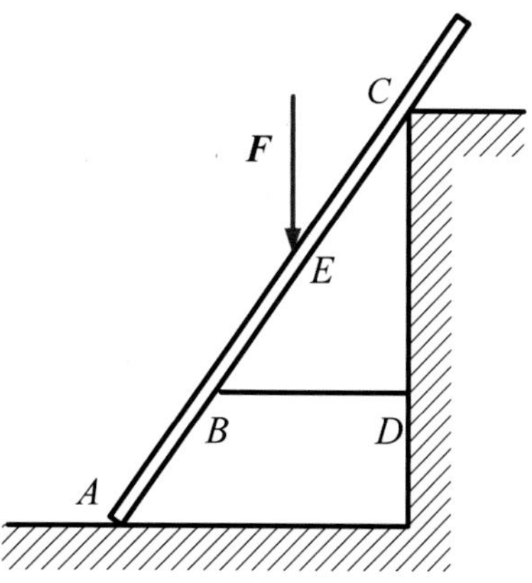

Fig. 3.27 A bar lying at a wall corner

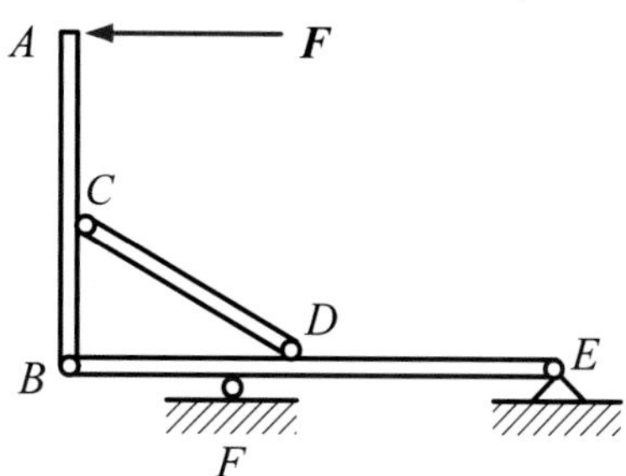

Fig. 3.28 A system including three bars

Answers:

3.1

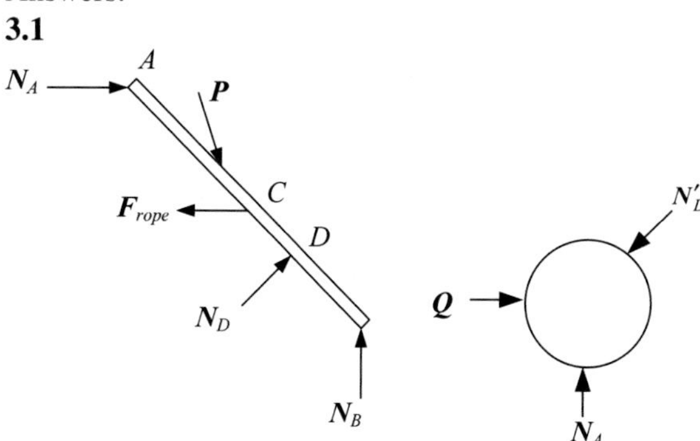

3.2

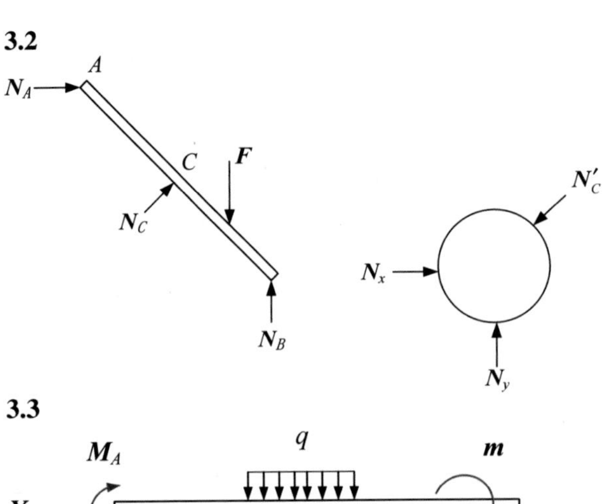
A
N_A
C
F
N_C
N_B
N'_C
N_x
N_y

3.3

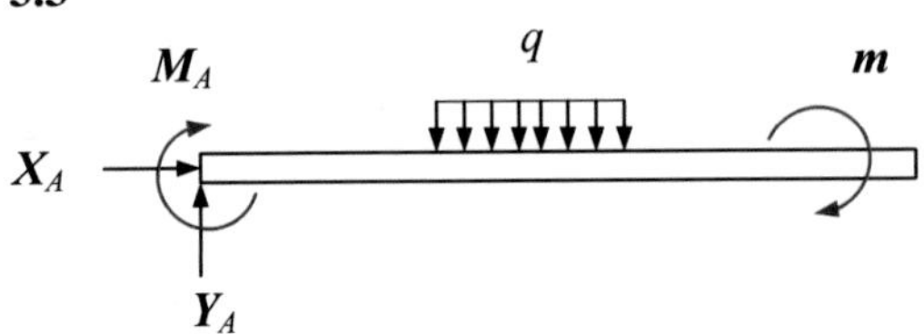
M_A
q
m
X_A
Y_A

3.4

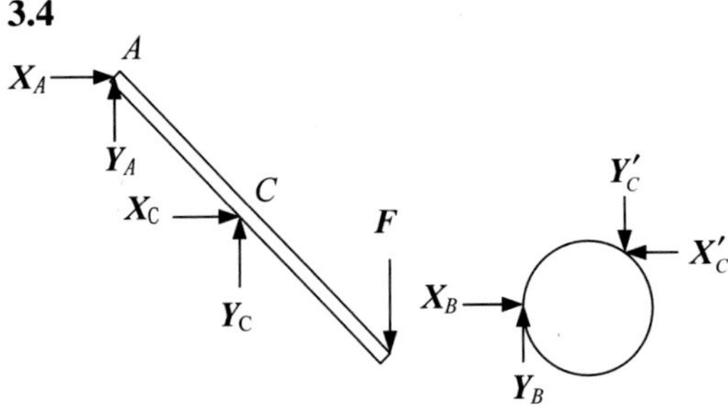
A
X_A
Y_A
C
X_C
Y_C
F
Y'_C
X'_C
X_B
Y_B

3.5

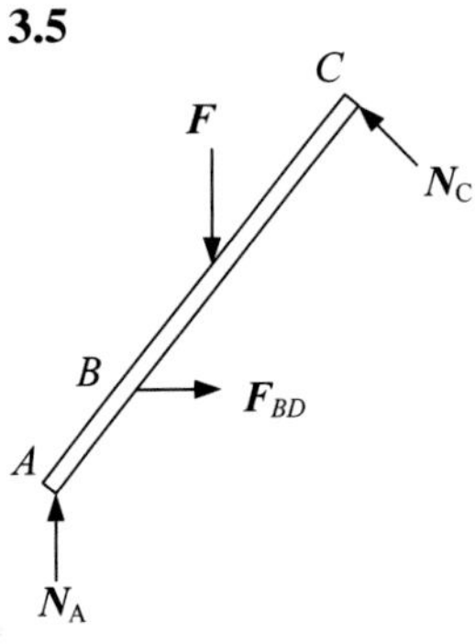
C
F
N_C
B
F_{BD}
A
N_A

3.6

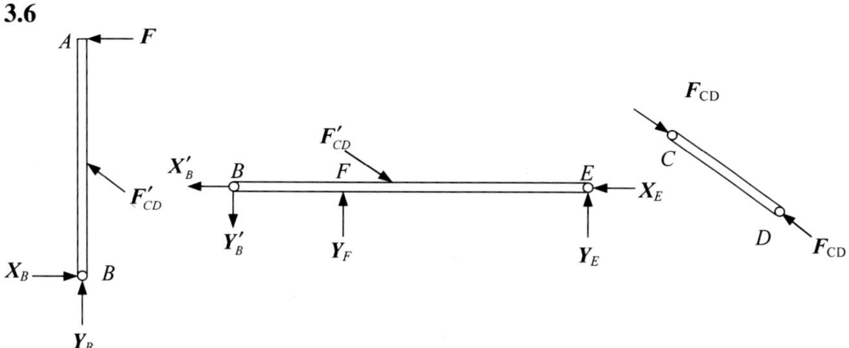

Chapter 4
Simplification of a Force Group

Abstract In this chapter, we will learn how to simplify a force group. First, the concept of planar couple is given, and then the theorem of parallel translation of a force is introduced. Based on these theories, a complex force system can be simplified, and can also be further simplified.

Keywords Planar couple · Parallel translation of a force · Simplification of a force system · Further simplification of a force system

Generally speaking, a rigid body endures a group of forces, which is difficult to analyze in practice. For example, when an airplane is flying in the sky, there are several forces acting on its body. In this situation, how to steer the direction of this airplane? It is imperative to judge the resultant force of this system. This procedure is termed as simplification or reduction of a force group.

4.1 Planar Couple

We herein first introduce the concept of couple. A couple includes two parallel forces, with the same magnitudes, different directions, and the two forces have a distance between them. For example, we can control the rotation of the steering wheel to drive a car, where we have applied two forces constituting in one couple. Therefore, the steering wheel can rotate under the action of these special forces. In fact, the physical mechanism of the couple is to cause the rotation of the rigid body.

We mainly investigate the planar couple, as shown in Fig. 4.1. The couple can be expressed as $(\boldsymbol{F}, \boldsymbol{F}')$. Since the couple includes two forces, it has a moment, which is the summation of the two moments caused by the two forces. The couple moment is expressed as $M = Fa$, where a is the distance between the two forces.

For the planar couple, we dictate that if the rotation direction is anticlockwise, its sign is positive; otherwise, it is negative, which is shown in Fig. 4.2. As the couple represents the rotation of the object, it can be moved to any point in the plane, which cannot change the state of the system.

J. Liu, *Lecture Notes on Theoretical Mechanics*,
https://doi.org/10.1007/978-981-13-8035-8_4

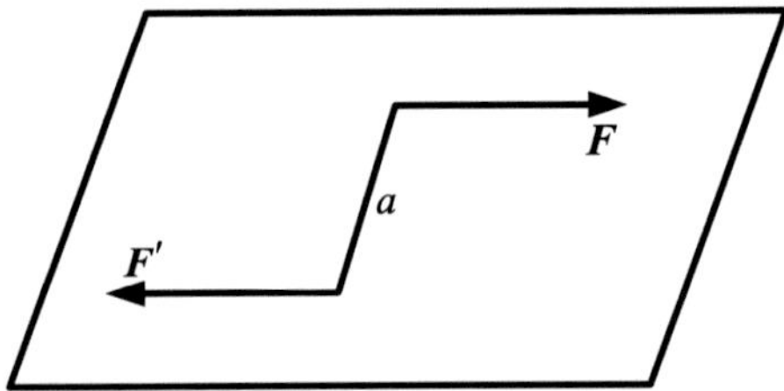

Fig. 4.1 A couple

Fig. 4.2 Direction of the couple

4.2 Theorem of Parallel Translation of a Force

To proceed, we first introduce a fundamental theory: the theorem of parallel translation of a force. The content of the theorem can be formulated as

A force acting on a rigid body can be moved parallel to its line of action to any point of the body, if we add a couple with a moment equal to the moment of the force about the point, to which it is translated.

This theorem can be proved as schematized in Fig. 4.3. If a force $\boldsymbol{F}$ is acting at point P, we can add an equilibrium force group at point Q, as shown in Fig. 4.3. This equilibrium force group consists of two forces, i.e., one is parallel to $\boldsymbol{F}$, and the other is opposite to the direction of $\boldsymbol{F}$. Both of the two forces have the same magnitude as that of the force $\boldsymbol{F}$. There is a distance a between the two parallel vectors in Fig. 4.3. Clearly, the force $\boldsymbol{F}$ at point P and force $\boldsymbol{F}'$ at point Q constitute in a couple. The final state is that there remains a force acting at point Q and a couple, and the moment of the couple $M = Fa$.

This theorem is quite salient and useful in practice. For example, as shown in Fig. 4.4, a column is experiencing an external force, whose line of action deviates from the axis of the column.

From the life experience, it is well known that the column will pierce into the ground substrate and rotate at the bottom end. In use of the above theorem, the secret can be easily disclosed. If the external force is moved parallel to the axis line of the column, there remain a force and a couple. The force can push the column into the substrate, and the couple can cause rotation of the column.

The second example is a small boat, where there is a person utilizing two oars to push the boat. As shown in Fig. 4.5, if the two oars are symmetrically pushing the

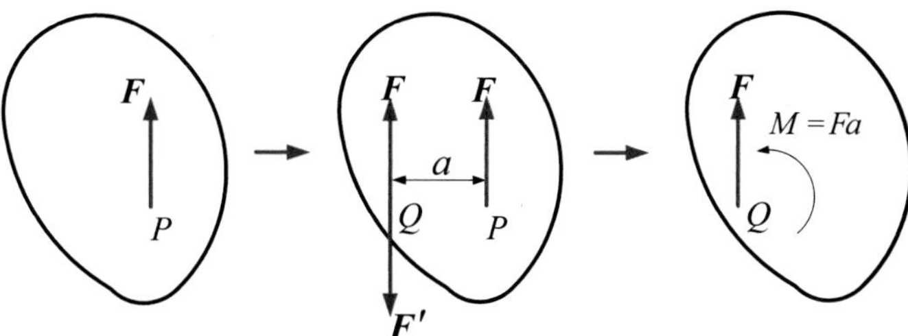

Fig. 4.3 Schematic of the parallel translation of a force

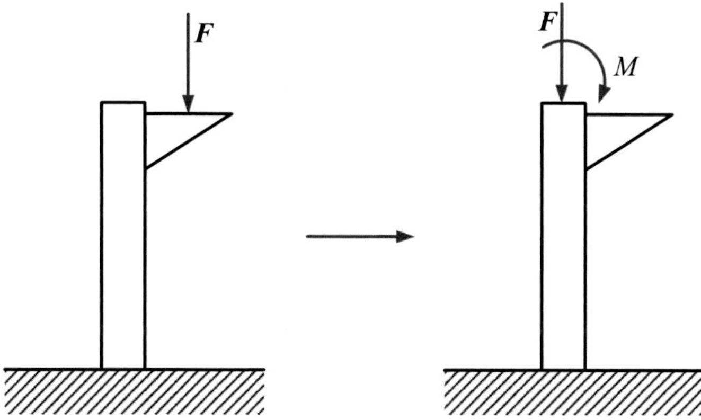

Fig. 4.4 A force acting on a column

Fig. 4.5 Only one oar is acting on the boat

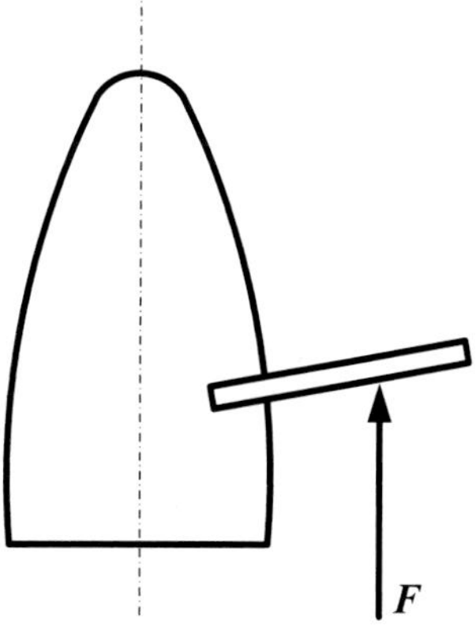

water, the boat can move forward. However, if only one oar is used, then the boat can move and rotate. The mechanism is similar to the last example. If the force from the reaction of water moves to the axis of the boat, then we can get one force and one couple, which cause the movement and rotation, respectively.

4.3 Simplification of a Force System

As shown in Fig. 4.6, there are a lot of forces acting on a rigid body, i.e., $\boldsymbol{F}_1$, $\boldsymbol{F}_2$, $\boldsymbol{F}_3$, …, $\boldsymbol{F}_n$. So many forces make the force system quite complex, and it is necessary to simplify the system. By using the theorem of parallel translation of a force, each force can be parallelly moved to one point O, accompanied by the additional couples $\boldsymbol{M}_1$, $\boldsymbol{M}_2$, $\boldsymbol{M}_3$, …, $\boldsymbol{M}_n$. All the forces are crossed at one point, which constitute in a concurrent force group. In addition, all the couples constitute one couple group. The concurrent force system can be simplified to one force acting at point O, which reads

$$\boldsymbol{R} = \Sigma \boldsymbol{F}_i.$$

This force is also referred to as the principal vector, as it's the summation of all the force vectors. All the couples can lead to one total couple, whose moment is

$$\boldsymbol{M}_O = \Sigma \boldsymbol{M}_O(\boldsymbol{F}_i).$$

This moment is often termed as the principal moment, which is dependent on the simplification center O.

However, we herein only concentrate on the general coplanar force system. In this case, the conclusion is that, for a general coplanar force system, the final simplification results are one force and one couple. The force is equal to the principal vector, which is passing through the simplification center O. The moment of the couple is equal to the principal moment. That is, the force is expressed as

$$\boldsymbol{R} = \Sigma \boldsymbol{F}_i.$$

and the couple is written as

$$M_O = \Sigma M_O(\boldsymbol{F}_i).$$

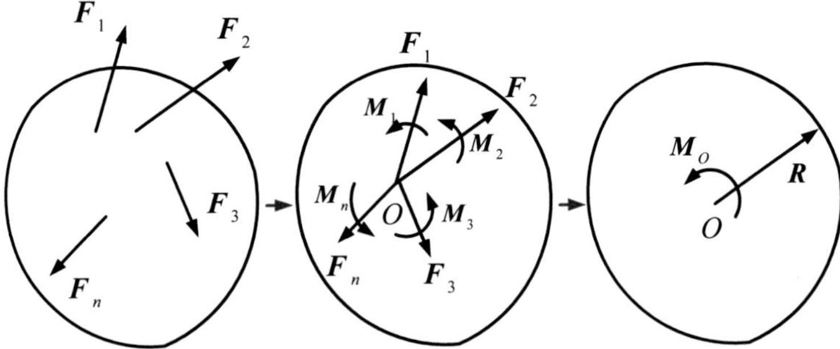

Fig. 4.6 Schematic of simplification of a force group

4.4 Further Simplification of a Force System

For the practical cases, we should be concerned with the actual simplification of a general coplanar force group. There are mainly four possible cases for the simplified result of one force and one couple:

(1) $\boldsymbol{R} = 0$ and $M_O = 0$, corresponding to the equilibrium state of the rigid body, which will be discussed later. In this case, the system is one zero-force system.
(2) $\boldsymbol{R} \neq 0$ and $M_O = 0$, corresponding to one resultant force, which passes through the simplification center O.
(3) $\boldsymbol{R} = 0$ and $M_O \neq 0$, corresponding to one couple, whose magnitude is equal to the principal moment.
(4) $\boldsymbol{R} \neq 0$ and $M_O \neq 0$, corresponding to one resultant force. This can be further proved as follows.

The couple can be displaced with two forces $\boldsymbol{R}$ and $\boldsymbol{R}' = -\boldsymbol{R}$, as shown in Fig. 4.7. The distance between the two parallel forces is d, and one has the relation $M_O = Rd$. At point O, the forces $\boldsymbol{R}$ and $\boldsymbol{R}'$ constitute in an equilibrium force system, and can be removed from the rigid body. Therefore, there remains only one force, which is the resultant force. The distance between the current resultant force and the initial force is d, which is expressed as

$$d = \left| \frac{M_O}{R} \right|.$$

Example 1: As shown in Fig. 4.8, a square is experiencing three forces. The side length of the square is a.
Answer: The principal vector is

$$\boldsymbol{R} = \boldsymbol{P}.$$

(Normally, we first solve R_x and R_y).

We select point O as the simplification center. The principal moment is

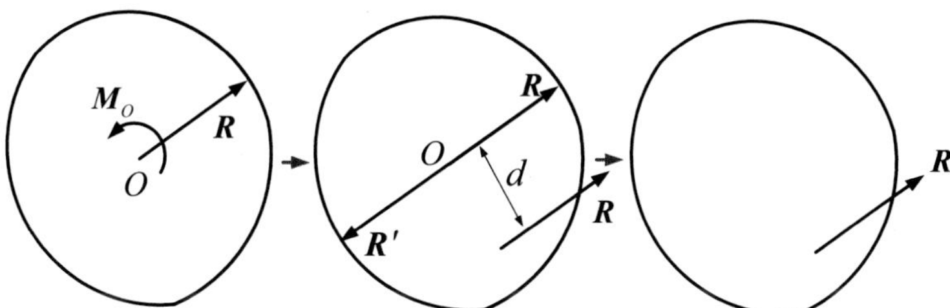

Fig. 4.7 Further simplification of one force and one couple

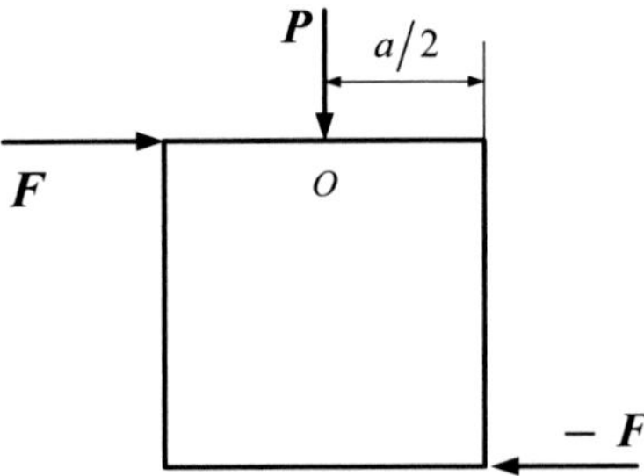

Fig. 4.8 One square enduring three forces

$$M_O = -Fa,$$

where the minus means the clockwise direction of the couple.

The distance is

$$d = \frac{M_O}{P} = \frac{Fa}{P}.$$

Especially, when $F = P$, $d = a$. The final result is schematized in Fig. 4.9.

Example 2: An equilateral triangle experiencing three forces, as schematized in Fig. 4.10. The side length of the triangle is a. The magnitudes of the three forces are the same, which are all F.

Answer: The principal vector is

$$R_x = F_2 - F_1 \cos 60^\circ - F_3 \cos 60^\circ = 0,$$
$$R_y = F_3 \sin 60^\circ - F_1 \sin 60^\circ = 0.$$

(Normally, we first solve R_x and R_y).

We select one apex as the simplification center. The principal moment is

$$M_O = \frac{\sqrt{3}}{2} Fa.$$

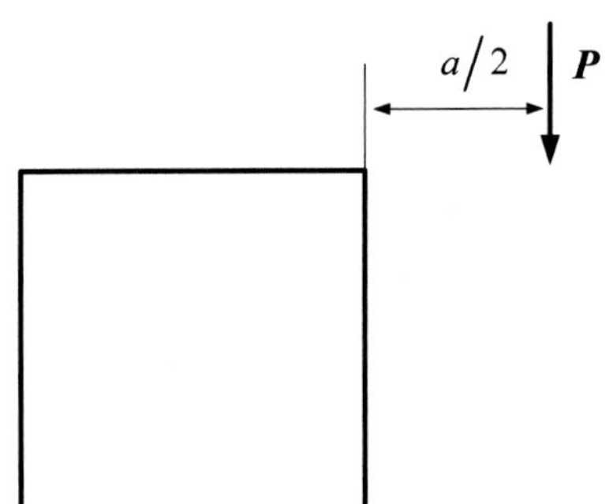

Fig. 4.9 Final simplification result

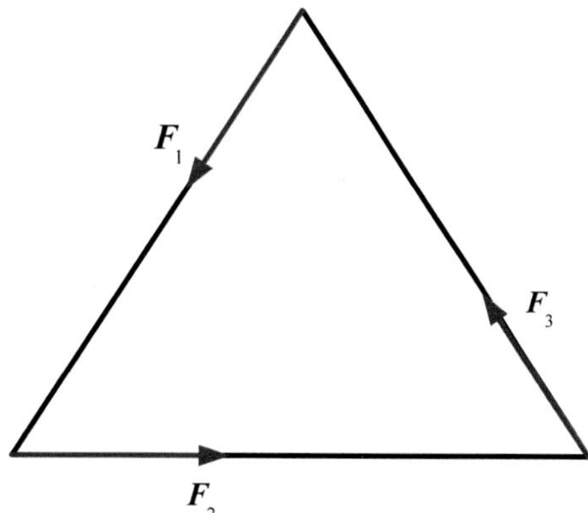

Fig. 4.10 One equilateral triangle enduring three forces

Therefore, the final simplification result is one couple, whose magnitude is $\frac{\sqrt{3}}{2}Fa$.

For the fixed end constraint or the clamped end shown in Fig. 4.11a, at point A, the beam cannot move and rotate. If we release the constraints, there is a group of forces acting at this end, as depicted in Fig. 4.11b. According to the simplification of the force group, we can get one force and one moment, as shown in Fig. 4.11c. For the purpose of convenient analysis, the force can be further decomposed into two forces $\boldsymbol{X}_A$ and $\boldsymbol{Y}_A$, as shown in Fig. 4.11d.

Finally, as conclusion which should be remembered, we will present the resultant forces for the distributed forces. As shown in Fig. 4.12, the resultant force for a

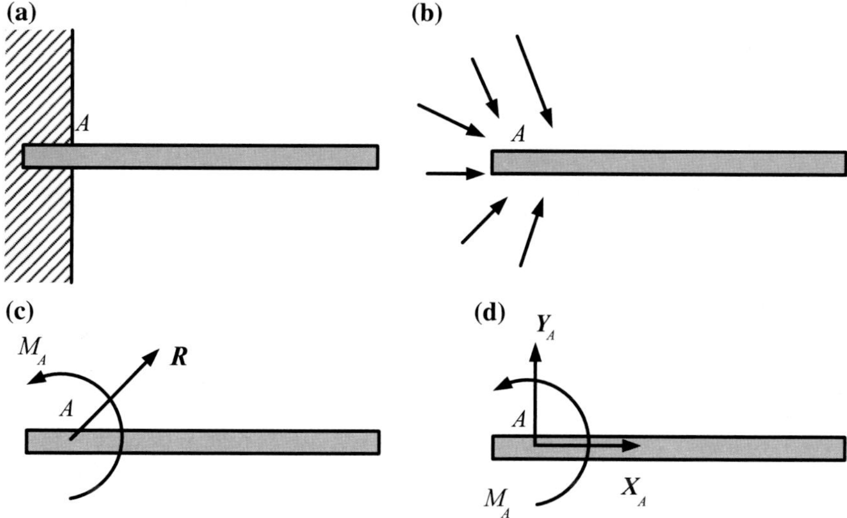

Fig. 4.11 Fixed end constraint

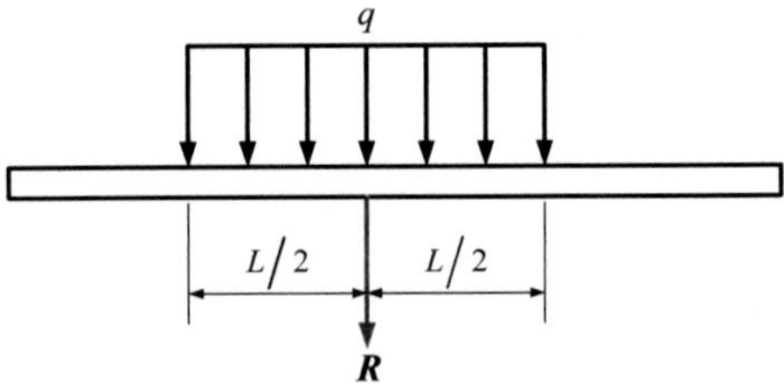

Fig. 4.12 Uniformly distributed force

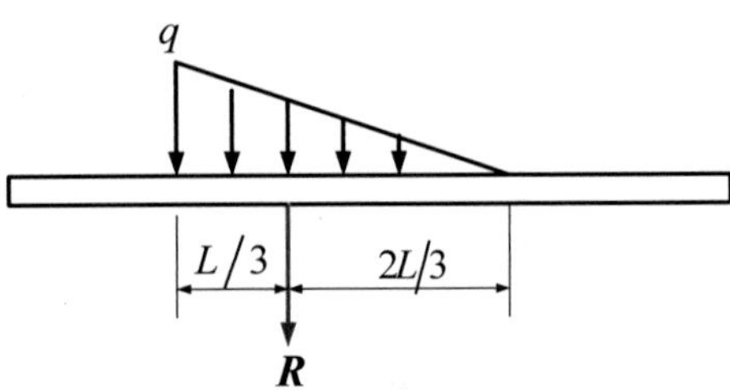

Fig. 4.13 Triangular distributed force

uniformly distributed force is located at the center of the spanning length of the distributed force.

For a triangular distributed force, the result force is located at the position which can separate the total distribution length of the force into 1/3 and 2/3, respectively. The force is displayed in Fig. 4.13. These conclusions will be explained in the later content.

Summary

In this chapter, we have learned how to simplify a force group. We first need master the concept of planar couple, and the theorem of parallel translation of a force. Based on these theories, a complex force system can be simplified, and can also be further simplified.

Exercises

4.1 As shown in Fig. 4.14, there is a triangle *ABC* with a right angle, and the angle $A = 30°$. The length of *BC* equals *a*. There are four forces applied at the three points, with the same magnitude *P*. Please give the final result of the simplification on this force group.

4.2 As shown in Fig. 4.15, there is a triangle *ABC* with a right angle, and the angle $A = 30°$. The length of *BC* equals *a*. There are two forces applied at the point *A* and *B*, respectively. The magnitude for these two forces is $P = qa$, and there is a uniformly distributed force *q* applied at *AC* side. Please give the final result of the simplification on this force group.

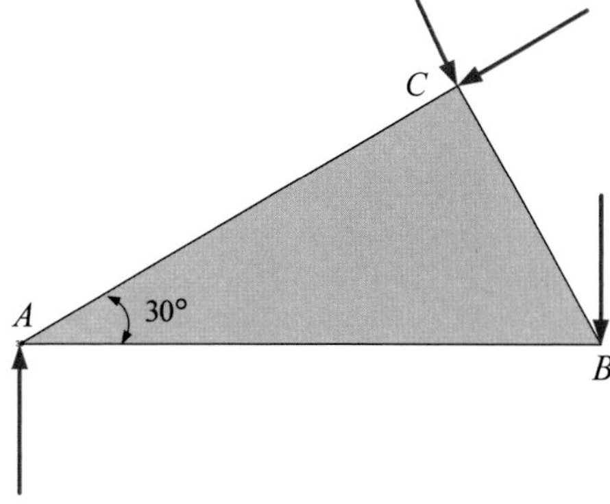

Fig. 4.14 A triangle force group with concentrated forces

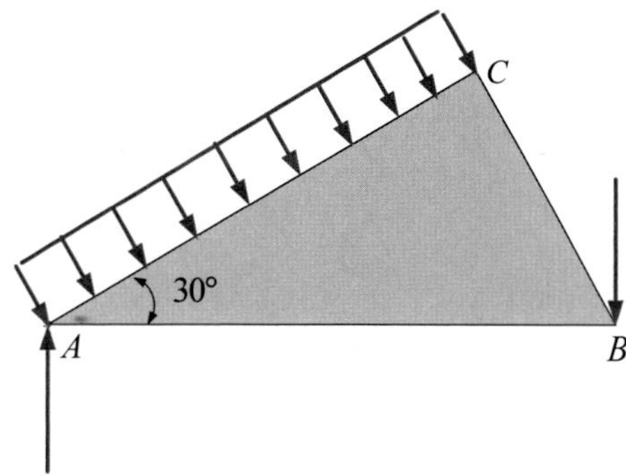

Fig. 4.15 A triangle force group with one distributed force and two concentrated forces

Answers

4.1 Select point A as the simplification center, and one has:
$F_x = \frac{1}{2}P - \frac{\sqrt{3}}{2}P,\ F_y = \frac{\sqrt{3}}{2}P - \frac{1}{2}P,\ M_A = -\left(2 + \sqrt{3}\right)Pa$

4.2 Select point A as the simplification center, and one has:
$F_x = \frac{3}{2}qa,\ F_y = -\frac{\sqrt{3}}{2}qa,\ M_A = -\frac{7}{2}qa.$

Chapter 5
Equilibrium of the General Coplanar Force Group

Abstract In this chapter, we will learn how to write down the equilibrium equations for a general coplanar force group. We mainly introduce two cases, i.e., the first one is on the one rigid body system and the other is on the rigid multi-body system, which is more complicated.

Keywords General coplanar force group · Equilibrium equation · One rigid body system · Rigid multi-body system

In a lot of engineering applications, when all the forces acting on the rigid body are located in the same plane, and their lines of action are not passing through the same point, this case is called the general coplanar force system. For example, as shown in Fig. 5.1, the first structure is enduring one force $\boldsymbol{P}$, one couple M, and the reaction forces $\boldsymbol{F}_x$, $\boldsymbol{F}_y$, and $\boldsymbol{F}_N$. The beam shown in Fig. 5.1 is experiencing one load $\boldsymbol{P}$, the gravity $\boldsymbol{Q}$, and the reaction forces $\boldsymbol{F}_{xA}$, $\boldsymbol{F}_{yA}$, and $\boldsymbol{F}_B$. These two examples are both general coplanar force groups.

5.1 Equilibrium Equations for One Rigid Body System

Let us continue to discuss the equilibrium conditions of the general coplanar force group. As mentioned in Chap. 4, the equilibrium conditions are that the principal vector and principal moment are both zeroes:

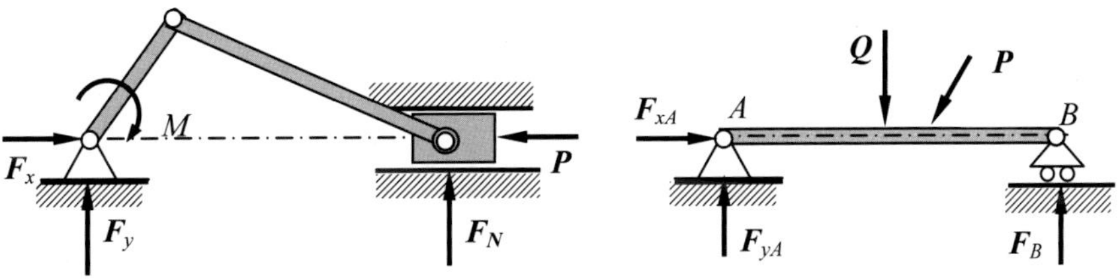

Fig. 5.1 Two examples of general coplanar force group

J. Liu, *Lecture Notes on Theoretical Mechanics*,
https://doi.org/10.1007/978-981-13-8035-8_5

$$\boldsymbol{R} = 0,$$
$$M = 0.$$

In the Cartesian coordinate system, the above formulas are expressed as

$$\Sigma X = 0,$$
$$\Sigma Y = 0,$$
$$\Sigma M_A = 0.$$

This group of formulas includes two equations on force, and one equation on moment, so it is normally named as one moment format. In fact, there are also some other formats on the equilibrium conditions. For example, the two-moment format can be written as

$$\Sigma X = 0,$$
$$\Sigma M_A = 0,$$
$$\Sigma M_B = 0.$$

The three-moment format can be expressed as

$$\Sigma M_A = 0,$$
$$\Sigma M_B = 0,$$
$$\Sigma M_C = 0.$$

For the special cases, for instance, the coplanar concurrent force system, the equations can degenerate into two equations

$$\Sigma X = 0,$$
$$\Sigma Y = 0.$$

For the couple system, the equilibrium equation is

$$\Sigma M = 0.$$

Here, we should stress the steps to solve the general coplanar force system:

(1) Select the object, and isolate it from the neighbors;
(2) Perform the force analysis, including the active forces and reactive forces;
(3) Lay out the equilibrium equations;
(4) Give the final answers.

Example 1 As shown in Fig. 5.2, two bars AC and BC are hinged together at point C, where there is a concentrated force $\boldsymbol{P}$. The angle $\theta = 30°$. Please solve the internal forces of the two bars.

Answer: This is an example of the concurrent force system. The two bars are both two-force bars. We select point C as an object, and it endures three forces, i.e., the concentrated force $\boldsymbol{P}$, and reactive forces from the two bars. The free body diagram is displayed in Fig. 5.3.

The equilibrium equations are

$$\Sigma X = 0, S_{BC} - S_{AC}\cos\theta = 0$$
$$\Sigma Y = 0, S_{AC}\sin\theta - P = 0$$

(This format is very important, and it should always be followed!)

The solutions are $S_{AC} = 2P$ and $S_{BC} = \sqrt{3}P$.

Example 2 As shown in Fig. 5.4, the left end of beam AB is hinged at point A, and point B is a roller support. The length of the beam is $4a$, with the gravity being $\boldsymbol{P}$. Point C is the midpoint of the beam. There are two distributed forces, i.e., one uniformly distributed force and one triangular distributed force. Please give the reactive forces at point A and B.

Answer: This example is also a general coplanar force group. We select the beam AB as an object, which is shown in Fig. 5.5. The beam AB is subjected to the active force $\boldsymbol{P}$, the distributed forces, and the reactive forces.

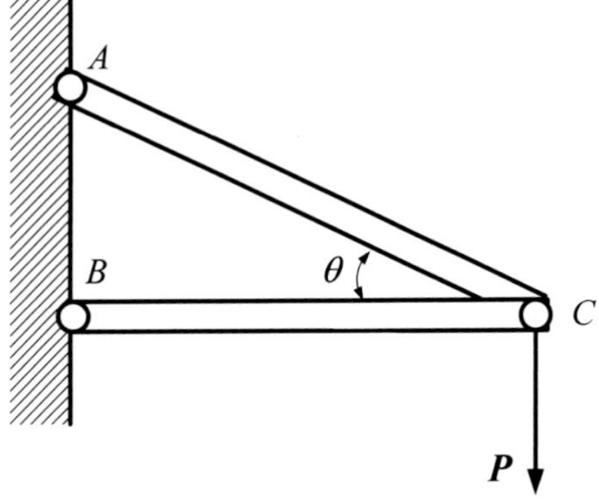

Fig. 5.2 Two bars hinged together subject to a concentrated force

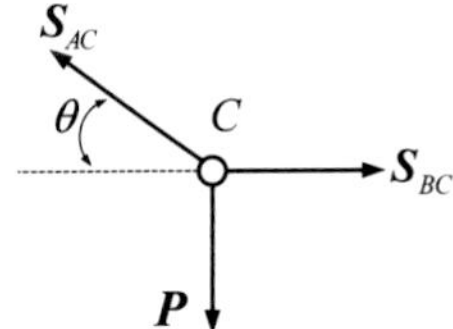

Fig. 5.3 Point C enduring three forces

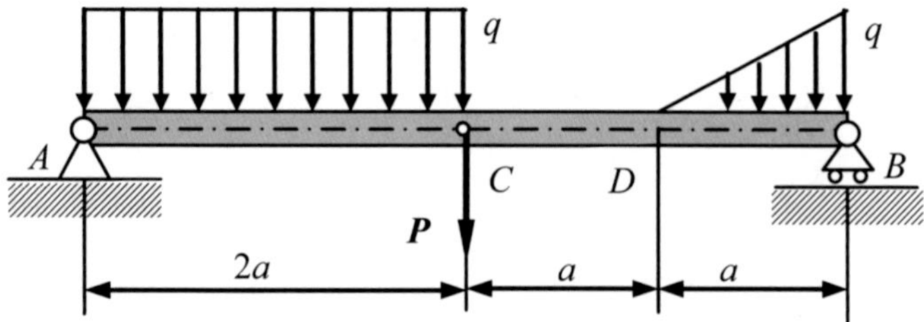

Fig. 5.4 A beam enduring several forces

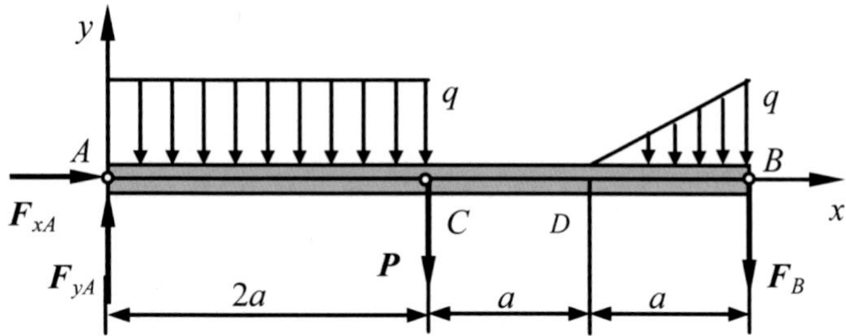

Fig. 5.5 Free body diagram

The equilibrium equations are

$$\Sigma X = 0, F_{xA} = 0$$

$$\Sigma M_A = 0, F_B \cdot 4a + P \cdot 2a + 2qa \cdot a + \frac{qa}{2} \cdot \frac{11a}{3} = 0$$

$$\Sigma M_B = 0, F_{yA} \cdot 4a - P \cdot 2a - 2qa \cdot 3a - \frac{qa}{2} \cdot \frac{a}{3} = 0$$

The answers are $F_{xA} = 0, F_{yA} = \frac{P}{2} + \frac{37qa}{24}$ and $F_B = -\frac{P}{2} - \frac{23qa}{24}$.

We can also verify this result, and substitute the above solutions into the equation $\Sigma Y = 0$. The equation is $\Sigma Y = F_{yA} - F_B - 2qa - \frac{1}{2}qa - P = 0$.

Example 3 As shown in Fig. 5.6a, a slender bar *AB* is lying on two surfaces which are perpendicular to each other. The gravity of the bar is ***P***, and gravitational center *C* is the middle point of line *AB*. The inclination angle in the figure is α. Please give the value of θ when the bar is in equilibrium, and the reactive forces of point *A* and *B*.

Answer: We select the bar *AB* as an object. The free body diagram is shown in Fig. 5.6b, and we have established a Cartesian coordinate system *o-xy*. Suppose the length of the bar is *l*.

The equilibrium equations are

$$\Sigma X = 0, F_A - P \cos \alpha = 0$$

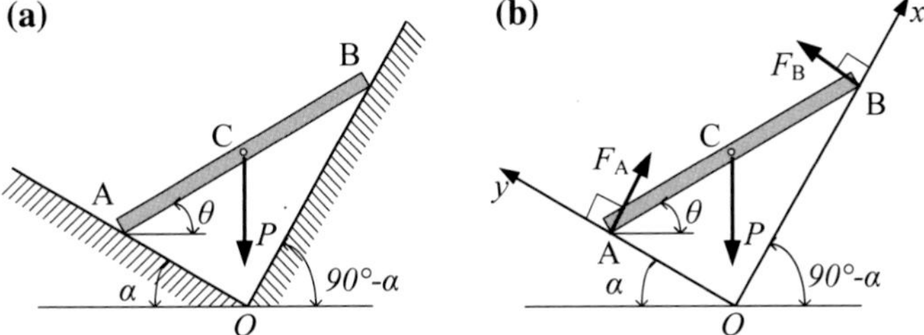

Fig. 5.6 A bar is lying between two surfaces

$$\Sigma Y = 0, F_B - P\sin\alpha = 0$$

$$\Sigma M_A = 0, F_B l\sin(\alpha+\theta) - P\frac{l}{2}\cos\theta = 0$$

After some derivations, one has

$$2\sin\alpha(\sin\alpha\cos\theta + \cos\alpha\sin\theta) - \cos\theta = 0.$$

That is

$$\sin 2\alpha\sin\theta - \cos 2\alpha\cos\theta = 0$$

and

$$\tan\theta = \cot 2\alpha = \tan(90^\circ - 2\alpha).$$

Then we have $\theta = 90^\circ - 2\alpha$.
From this result, we can see that when $\alpha < 45^\circ, \theta > 0$. When $\alpha > 45^\circ, \theta < 0$.

5.2 Rigid Multi-body System

The previous discussions are only confined to one rigid body. However, in most cases, we will face a system, which is assembled by several rigid bodies via constraints. This system is named as "rigid multi-body system". There are two types of forces for this kind of system. One is the external force, which comes from the other objects outside the system, and the other is internal force, which is the interaction force among the rigid bodies in the system. For the rigid multi-body system, we have the following rules:

(1) When the whole system is in equilibrium, every rigid body in the system must be in equilibrium.

(2) We can select a portion of the system as an object, or can choose the whole system to investigate.
(3) When we consider the total system, the internal forces among the subsystems do not appear; and when we analyze one subsystem, it endures the action forces from the other subsystems.
(4) The action and reaction forces always appear together, with the opposite directions.
(5) For a system including n rigid bodies, the number of the individual objects we can choose is n.

It can be seen that, for a rigid multi-body system including n rigid bodies, we can write down $3n$ equilibrium equations totally. If the number of the unknowns is equal to that of the equilibrium equations, all of these unknowns can be solved, and the case is called determinate problem. And vice versa, if the number of the unknowns is bigger than that of the equilibrium equations, not all of these unknowns can be solved, and the case is called indeterminate problem. How to solve the indeterminate problem needs considering the deformation of the object, which is out of the range of Statics.

Example 4 As shown in Fig. 5.7, the structure in equilibrium includes one wheel I, a bar AB, and a punch B. Point A and B are both hinged constraints. $OA = R, AB = l$. The gravity effect on the objects is ignored. When OA is in the horizontal position, the force of the punch is $\boldsymbol{P}$. Please solve the following:

(1) The moment M acting on the wheel.
(2) The reaction forces at point O.
(3) The force of the bar AB.
(4) The lateral force of the punch to the orbit.

Answer: This is a rigid multi-body system in equilibrium.

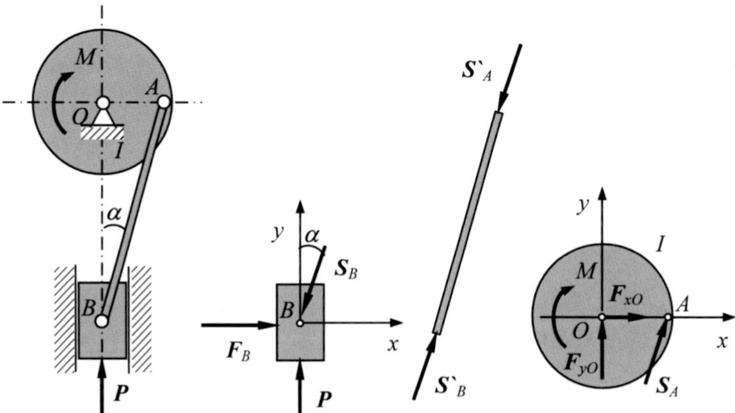

Fig. 5.7 The mechanical structure in equilibrium

(1) We select the punch B as an object. The free body diagram is shown in Fig. 5.7. The equilibrium equations are as follows:

$$\Sigma X = 0, F_B - S_B \sin\alpha = 0$$
$$\Sigma Y = 0, P - S_B \cos\alpha = 0$$

Therefore, one has $S_B = \frac{P}{\cos\alpha} = \frac{Pl}{\sqrt{l^2-R^2}}$. The sign is positive, so the assumed direction of S_B is right. We then further get $F_B = P\tan\alpha = \frac{PR}{\sqrt{l^2-R^2}}$. According to the Newton's third law, the lateral force of the punch to the orbit is $F'_B = F_B = P\tan\alpha$.

(2) AB is a two-force bar. We have $S_A = S'_A = S'_B = S_B = \frac{P}{\cos\alpha} = \frac{Pl}{\sqrt{l^2-R^2}}$.

(3) We select the wheel I as an object. This is a general coplanar force system. The equilibrium equations read

$$\Sigma X = 0, F_{xO} + S_A \sin\alpha = 0$$
$$\Sigma Y = 0, F_{yO} + S_A \cos\alpha = 0$$
$$\Sigma M_O(F_i) = 0, S_A R\cos\alpha - M = 0$$

We finally obtain $F_{xO} = -P\tan\alpha = -\frac{PR}{\sqrt{l^2-R^2}}$, $F_{yO} = -P$, and $M = PR$. The minus means that the directions of the forces are opposite to the assumed ones.

Example 5 As shown in Fig. 5.8, a horizontal beam is composed of two portions, i.e., AC and CD, which are hinged at point C. The left end of the beam A is clamped, and point B is a roller support. The force $Q = 10$ kN, $P = 20$ kN, and the uniformly distributed force $p = 5$ kN/m. There is a linearly distributed load in the BD segment, and the maximum value at point D is $q = 6$ kN/m. Please give the reaction forces at point A, B, and C.

Answer: There are two rigid bodies in this system.

(1) We first select the segment CD to investigate, whose free body diagram is shown in Fig. 5.9. The equilibrium equations are

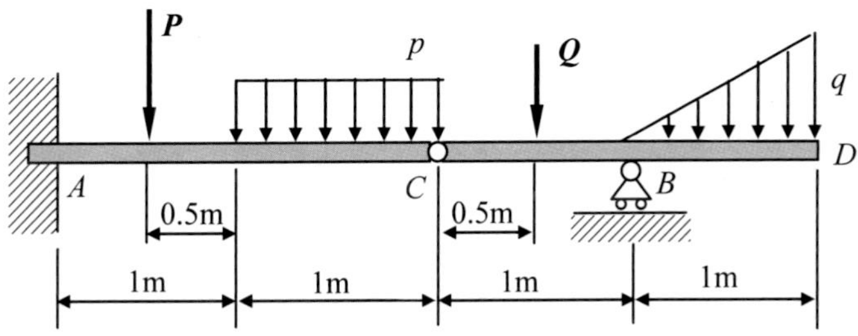

Fig. 5.8 The total forces and structures

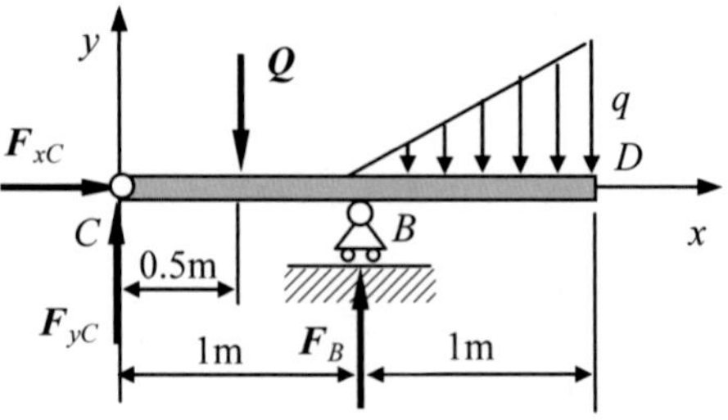

Fig. 5.9 The free body diagram of *CD*

$$\Sigma X = 0, F_{xC} = 0$$
$$\Sigma Y = 0, F_{yC} + F_B - Q - q \times 1 \times 1/2 = 0$$
$$\Sigma M_O(F_i) = 0, F_B \times 1 - Q \times 0.5 - \frac{1}{2}q \times 1 \times (1 + \frac{2}{3}) = 0.$$

From the above equations we have $F_{xC} = 0$, $F_{yC} = 3\,\text{kN}$, and $F_B = 10\,\text{kN}$.

(2) We then choose the segment of *AC* to analyze, whose free body diagram is shown in Fig. 5.10. The equilibrium equations are

$$\Sigma X = 0, F_{xA} - F'_{xC} = 0$$
$$\Sigma Y = 0, F_{yA} - F'_{yC} - P - p \times 1 = 0.$$

Finally, one has $F_{xA} = 0$, $F_{yA} = 28\,\text{kN}$, and $M_A = 23.5\,\text{kN m}$.

Example 6 In engineering applications, there are a lot of bars hinged together to form a structure, which is termed as truss. We ignore the gravity of each truss, and every truss is assumed to be a two-force bar. As shown in Fig. 5.11, there is a truss, enduring one concentrated force $P = 10$ kN at point *D*. Please solve the internal force for each bar.

Answer: We first select the whole system to investigate, whose free body diagram is shown in Fig. 5.11. The equilibrium equations are

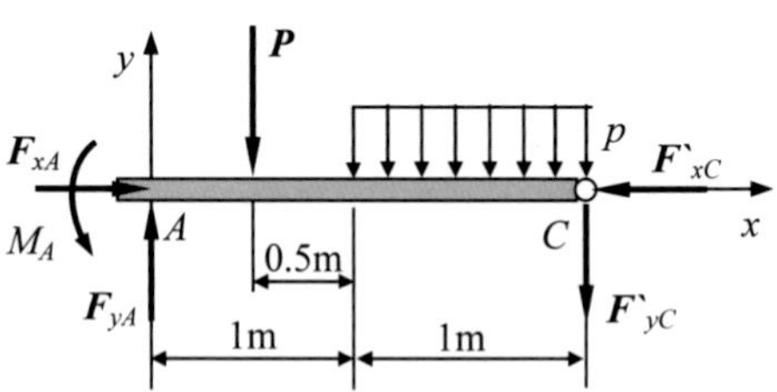

Fig. 5.10 The free body diagram of *AC*

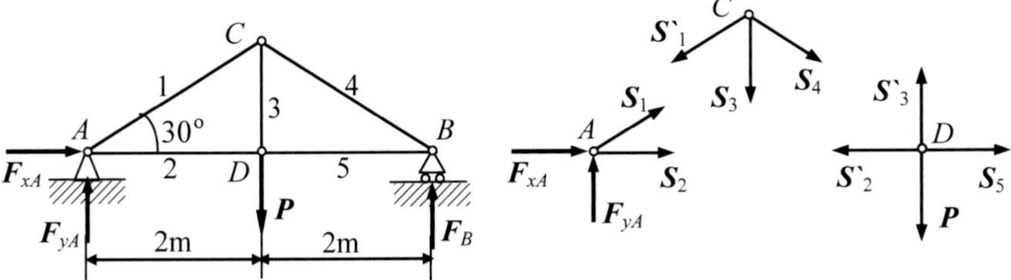

Fig. 5.11 A truss enduring one concentrated force

$$\begin{aligned}
&\Sigma X = 0, F_{xA} = 0 \\
&\Sigma Y = 0, F_{yA} + F_B - P = 0 \\
&\Sigma M_O(\boldsymbol{F}_i) = 0, F_B \times 4 - P \times 2 = 0.
\end{aligned}$$

Then we have $F_{xA} = 0$ and $F_B = F_{yA} = 5\,\text{kN}$.

We then analyze point A. This is a concurrent force system. The equilibrium equations are

$$\begin{aligned}
&\Sigma X = 0, F_{xA} + S_2 + S_1 \cos 30^\circ = 0 \\
&\Sigma Y = 0, F_{yA} + S_1 \sin 30^\circ = 0.
\end{aligned}$$

Then we have $S_1 = -10\,\text{kN}$ and $S_2 = 8.66\,\text{kN}$.

Next, we investigate point C. The equilibrium equations are

$$\begin{aligned}
&\Sigma X = 0, S_4 \cos 30^\circ - S_1' \cos 30^\circ = 0 \\
&\Sigma Y = 0, -S_3 - (S_4 + S_1') \sin 30^\circ = 0.
\end{aligned}$$

Then we have $S_4 = -10\,\text{kN}$ and $S_3 = 10\,\text{kN}$.

We select point D, and the equation is

$$\Sigma X = 0, S_5 - S_2' = 0.$$

The result is $S_2 = S_5 = 8.66\,\text{kN}$, $S_1 = S_4 = -10\,\text{kN}$, and $S_3 = 10\,\text{kN}$. The minus means the assumed direction is opposite to the actual one.

Summary

In this chapter, we have learned how to write down the equilibrium equations for a general coplanar force group, and then give solutions. We mainly concentrate on two cases, i.e., the first one is on the one rigid body system, and the other is on the rigid multi-body system, which is more complicated.

Exercises

5.1 As shown in Fig. 5.12, two bars are hinged at point A. A horizontal force is exerted at point D. The length $AB = AC = BC = 2AD$. Please give the reactive forces at point B and C.

5.2 Please give the reactive forces at point A and C, as shown in Fig. 5.13.

5.3 Please give the reactive forces at point A and C, as shown in Fig. 5.14.

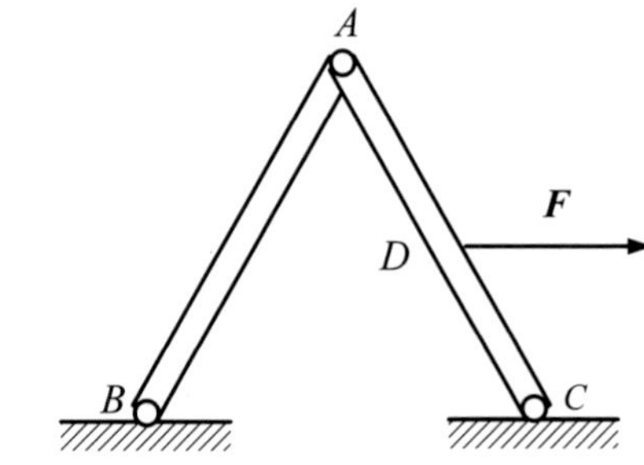

Fig. 5.12 Two bars hinged together

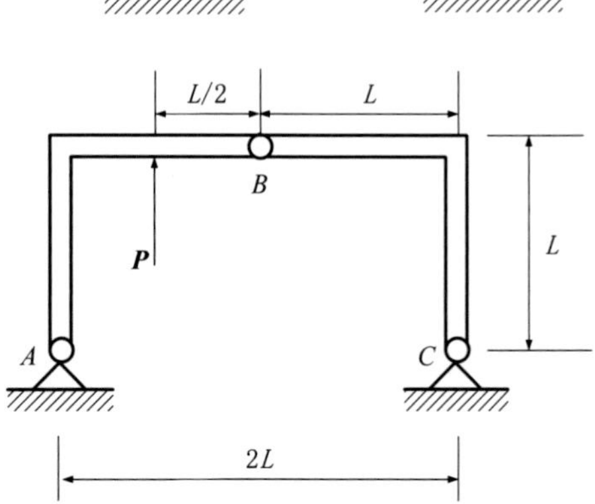

Fig. 5.13 Two bars hinged together

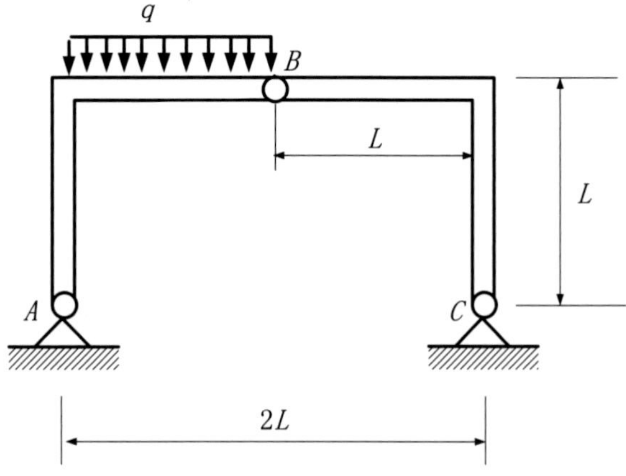

Fig. 5.14 Two curved bars hinged together

5.4 As shown in Fig. 5.15, there is a simply supported beam with the total length 8 m. A concentrated load $P = 20$ kN and a distributed load $q = 10$ kN/m are acting on the beam. Please calculate the reaction forces of point A and point B.

5.5 As shown in Fig. 5.16, two bars AB and BC are hinged together at point B, where a force $\boldsymbol{F}$ is applied at point D. The magnitude of the force $F = 11$ kN. The length $AD = 2L$, $AB = BC = 4L$. Please give the reactive forces at point A and C.

5.6 As shown in Fig. 5.17, a structure includes three bars. The length for each segment is marked in the figure. The weight for each bar is ignored, and the external forces are the uniformly distributed force q, concentrated force $\boldsymbol{P}$, and couple M. Please give the reactive loads at the fixed end.

5.7 As shown in Fig. 5.18, the bar AB and BE are linked at point B, and point A is the fixed end. The bar CD is linked with AB at point C, and linked with BE at point D. There is a uniformly distributed force acting on the DE segment, with the magnitude $q = 3$ kN/m. The lengths are $AB = BE = 2AC = 2BD = 4$ m. Please solve the reactive forces at point A and the internal force of the bar CD.

Answers

5.1 $X_B = -\frac{1}{4}F, Y_B = -\frac{\sqrt{3}}{4}F, X_C = -\frac{3}{4}F, Y_C = \frac{\sqrt{3}}{4}F$

5.2 $X_A = -\frac{1}{4}P, Y_A = -\frac{3}{4}P, X_C = \frac{1}{4}P, Y_C = -\frac{1}{4}P$

5.3 $X_A = \frac{1}{4}ql, Y_A = \frac{3}{4}ql, X_C = -\frac{1}{4}ql, Y_C = \frac{1}{4}ql$

5.4 $X_A = 0, Y_A = 0\,\text{kN}, Y_B = 35\,\text{kN}$

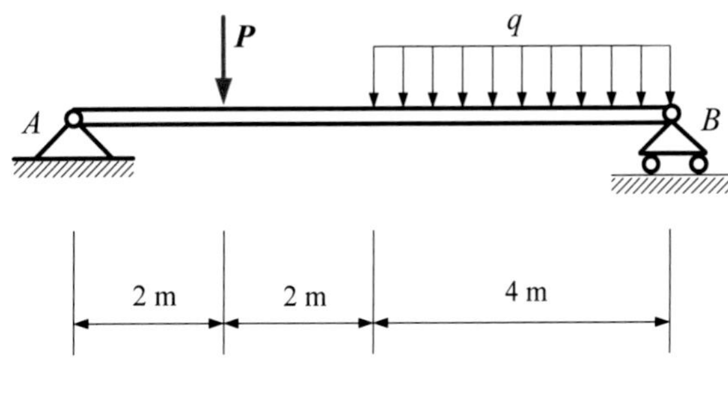

Fig. 5.15 One beam under external forces

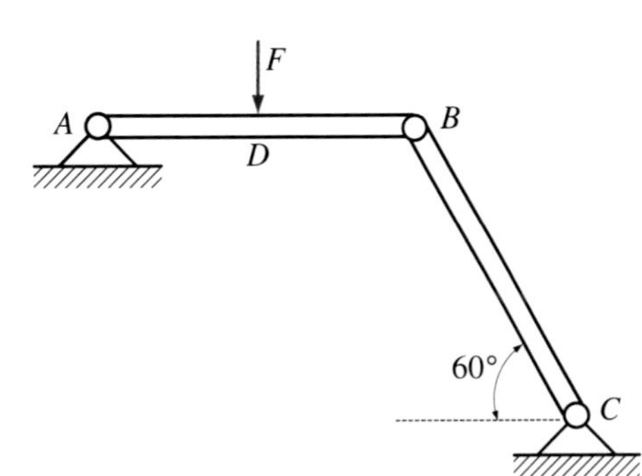

Fig. 5.16 Two bars hinged together

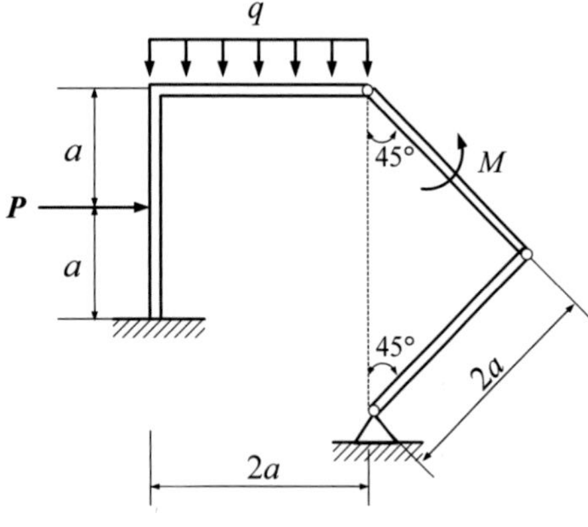

Fig. 5.17 A structure including three bars

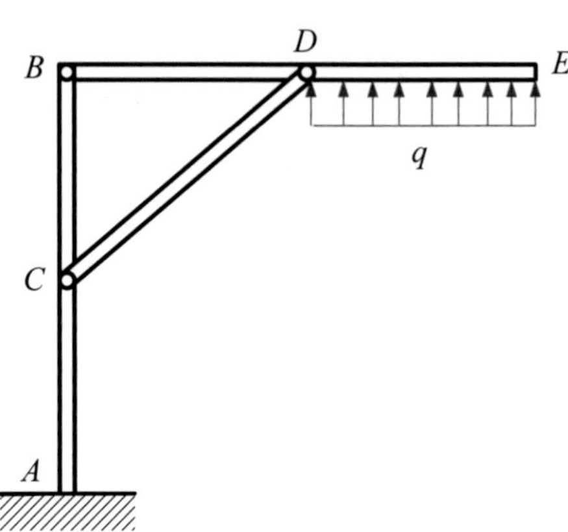

Fig. 5.18 A structure including three bars

5.5 $X_A = \frac{11\sqrt{3}}{6}\text{kN}, Y_A = \frac{11}{2}\text{kN}, X_C = -\frac{11\sqrt{3}}{6}\text{kN}, Y_C = \frac{11}{2}\text{kN}$

5.6 $X = \frac{\sqrt{2}M}{4a} - P, Y = \frac{\sqrt{2}M}{4a} + 2qa, M_{end} = Pa + 2qa^2$

5.7 $X_A = -6\,\text{kN}, Y_A = 0, M_A = 6\,\text{kN m}, F_{CD} = 9\sqrt{2}\,\text{kN}$

Chapter 6
Center of the Parallel Force Group

Abstract In this chapter, we will learn the knowledge on the center for a parallel force group. The content includes theorem of the resultant moment, center of the parallel force system, and gravitational center.

Keywords Theorem of the resultant moment · Parallel force system · Gravitational center · Centroid of the planar figure

6.1 Theorem of the Resultant Moment

We have already known the concept of mass center, and we mainly concentrate on the planar object without loss of the physical nature. In this case, the mass center of the volume degenerates into the centroid of the planar figure. In order to derive the formula of the centroid, we beforehand introduce the theorem of the resultant moment. From the mechanics viewpoint, the resultant force can be expressed as

$$\boldsymbol{R} = \sum \boldsymbol{F}_i.$$

The above expression means that the contribution of the resultant force to the moment is equal to the summation contributions of all the component forces. This leads to two aspects, i.e., the moment to an axis and the moment to a point, which can, respectively, be formulated as

$$\boldsymbol{M}_O(\boldsymbol{R}) = \sum \boldsymbol{M}_O(\boldsymbol{F}_i),$$

$$\begin{cases} M_x(\boldsymbol{R}) = \sum M_x(\boldsymbol{F}_i) \\ M_y(\boldsymbol{R}) = \sum M_y(\boldsymbol{F}_i). \\ M_z(\boldsymbol{R}) = \sum M_z(\boldsymbol{F}_i) \end{cases}$$

J. Liu, *Lecture Notes on Theoretical Mechanics*,
https://doi.org/10.1007/978-981-13-8035-8_6

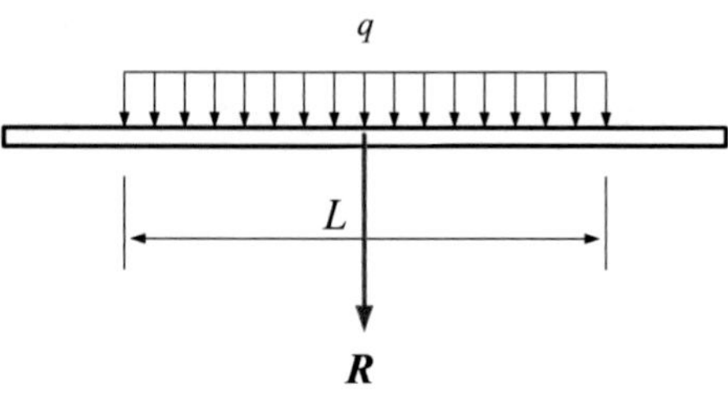

Fig. 6.1 Uniformly distributed force

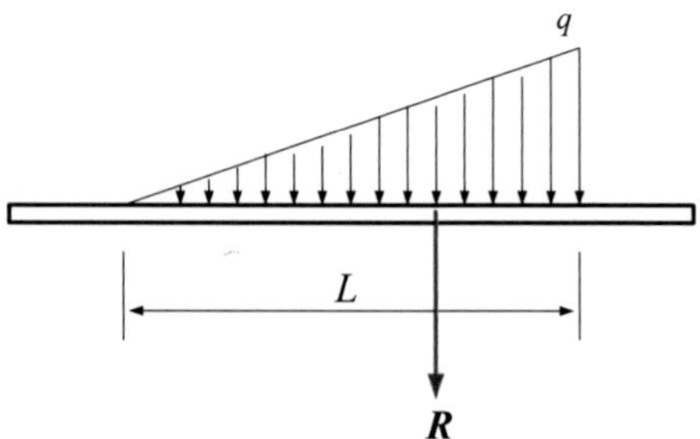

Fig. 6.2 Linearly distributed force

These theorems are grouped into the theorem of the resultant moment. We first look at two examples to use this theorem.

Example 1: As shown in Fig. 6.1, a uniformly distributed force is applied on a beam, with the span length of L, and the density of the force is q. Obviously, the resultant force is

$$R = qL,$$

and the moment center must be the middle point of the beam due to the symmetric configuration.

Example 2: As shown in Fig. 6.2, there is a linearly distributed force with the biggest magnitude q acting on the beam, with a span length of L. The resultant force is

$$R = \frac{qL}{2},$$

since the shadow area is a triangle. The theorem of the resultant moment leads to that the moment of the resultant force is equal to that of the distributed force. Then we can determine the position of the resultant force, which is near to the right end due to bigger magnitude of the force. It is proven that the resultant position has a distance of $L/3$ from the right end.

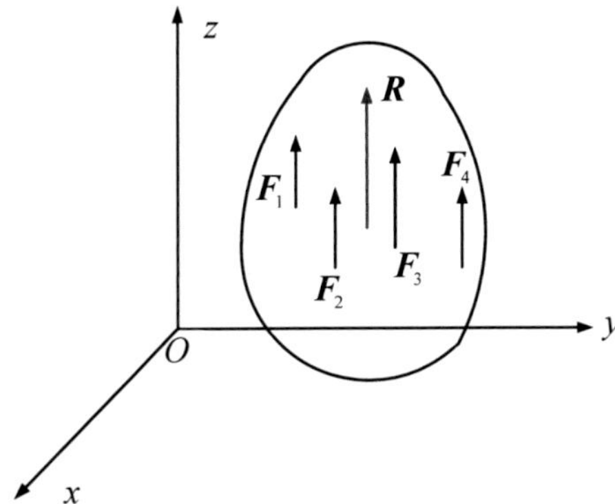

Fig. 6.3 Parallel force group

6.2 Center of the Parallel Force System

All the forces acting on the rigid body are parallel to each other, and this force group is called a parallel force group. In Fig. 6.3, all the forces are parallel to z-axis. The resultant force is

$$\boldsymbol{R} = \sum \boldsymbol{F}_i.$$

Each force has a distance to x-axis, which is denoted as $y_1, y_2, \ldots$. The distance to x-axis about the resultant force is called y_C, and the theorem of resultant moment leads to

$$Ry_C = \sum F_i y_i,$$

then one has

$$y_C = \frac{\sum F_i y_i}{R} = \frac{\sum F_i y_i}{\sum F_i}.$$

The similar derivations yield

$$x_C = \frac{\sum F_i x_i}{R} = \frac{\sum F_i x_i}{\sum F_i},$$

$$z_C = \frac{\sum F_i z_i}{R} = \frac{\sum F_i z_i}{\sum F_i},$$

The point (x_C, y_C, z_C) is called the center coordinate of the parallel force system, which is the action point of the resultant force.

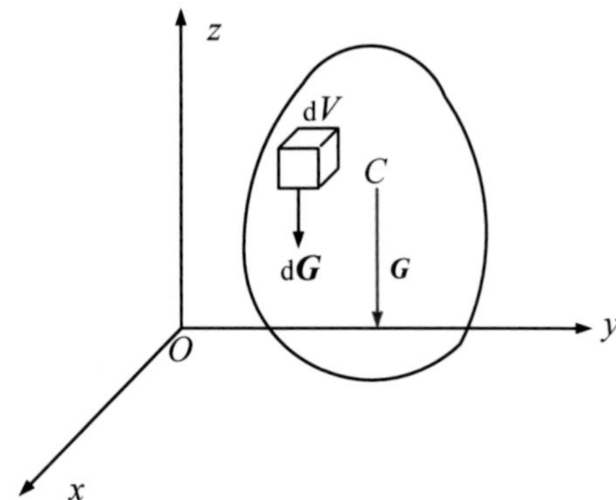

Fig. 6.4 Gravitational center

6.3 Gravitational Center

Let's consider the gravitational center of a continuous object, which deals with the integration around the whole volume. We select a micro-element with volume dV, and the summation effect of the micro-element constitutes in the whole body. The gravity is expressed as (Fig. 6.4)

$$G = \int_V \rho g \mathrm{d}V,$$

where ρ is the mass density of the body and g is the gravitational acceleration. The coordinate of the gravitational center is (x_C, y_C, z_C). According to the theorem of the resultant moment, one has

$$\begin{aligned} \mathrm{d}G \cdot y_C &= \rho g \mathrm{d}V \cdot y, \\ G y_C &= \int_V \rho g y \mathrm{d}V. \end{aligned}$$

Consequently, we have

$$y_C = \frac{\int_V \rho g y \mathrm{d}V}{G} = \frac{\int_V \rho y \mathrm{d}V}{\int_V \rho \mathrm{d}V}.$$

If the density is a constant, the above equation can be reduced to

$$y_C = \frac{\int_V y \mathrm{d}V}{V} = \frac{\int_V y \mathrm{d}V}{\int_V \mathrm{d}V}.$$

If we only consider the planar figure, the equation can further degenerate into

$$y_C = \frac{\int_A y \mathrm{d}A}{A} = \frac{\int_A y \mathrm{d}A}{\int_A \mathrm{d}A},$$

where A is the area of the planar figure. The similar derivations lead to

$$x_C = \frac{\int_A x \mathrm{d}A}{A} = \frac{\int_A x \mathrm{d}A}{\int_A \mathrm{d}A},$$

$$z_C = \frac{\int_A z \mathrm{d}A}{A} = \frac{\int_A z \mathrm{d}A}{\int_A \mathrm{d}A}.$$

For the planar figure, the gravitational center degenerates to the centroid.
For a single curve in one dimension, the centroid is

$$x_C = \frac{\int_l x \mathrm{d}l}{l} = \frac{\int_l x \mathrm{d}l}{\int_l \mathrm{d}l},$$

$$y_C = \frac{\int_l y \mathrm{d}l}{l} = \frac{\int_l y \mathrm{d}l}{\int_l \mathrm{d}l},$$

$$z_C = \frac{\int_l z \mathrm{d}l}{l} = \frac{\int_l z \mathrm{d}l}{\int_l \mathrm{d}l}.$$

For a discrete system including several planar figures ($i = 1, 2, \ldots, n$), with each area of the figure is A_i, and centroid C_i, the centroid coordinates can be expressed as

$$x_C = \frac{\sum A_i x_i}{A} = \frac{\sum A_i x_i}{\sum A_i},$$

$$y_C = \frac{\sum A_i y_i}{A} = \frac{\sum A_i y_i}{\sum A_i},$$

$$z_C = \frac{\sum A_i z_i}{A} = \frac{\sum A_i z_i}{\sum A_i}.$$

Example: A planar figure including two rectangles, with a symmetric configuration, as demonstrated in Fig. 6.5. Please name the centroid.
Answer: Due to symmetry, x_C= 50.

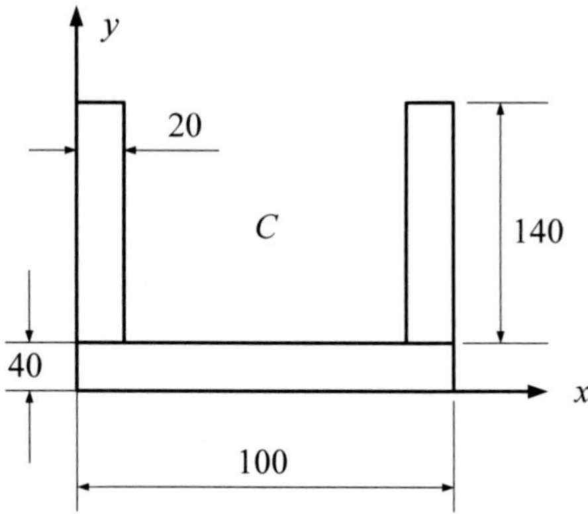

Fig. 6.5 A planar figure

$$y_{C1} = 110,\ y_{C2} = 110,\ y_{C3} = 20,\ A_1 = 2800 = A_2,\ A_3 = 4000$$

Then

$$y_C = \frac{A_1 y_{C1} + A_2 y_{C2} + A_3 y_{C3}}{A_1 + A_2 + A_3} = 72.5$$

We can also use the minus area method. We assume the planar figure includes two portions, one is the full rectangle, and the other is the blank rectangle.

$$y_{C1} = 90,\ y_{C2} = 110,\ A_1 = 18000,\ A_2 = -8400.$$

$$y_C = \frac{A_1 y_{C1} + A_2 y_{C2}}{A_1 + A_2} = 72.5$$

Summary

In this chapter, we must master the knowledge on the center for a parallel force group. The content includes theorem of the resultant moment, center of the parallel force system, and gravitational center. The most general examples are related to the centroid of a planar figure.

Exercises

6.1 Please name the position of the centroid as displayed in Fig. 6.6.

6.2 Please give the centroid of the planar figure shown in Fig. 6.7.

6.3 As shown in Fig. 6.8, a triangular area is removed from a circular area. The radius of the circle is 1. Please give the centroid of this planar figure.

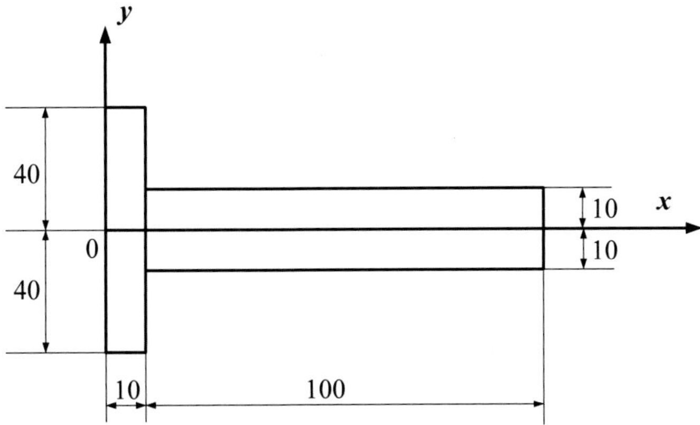

Fig. 6.6 A planar figure

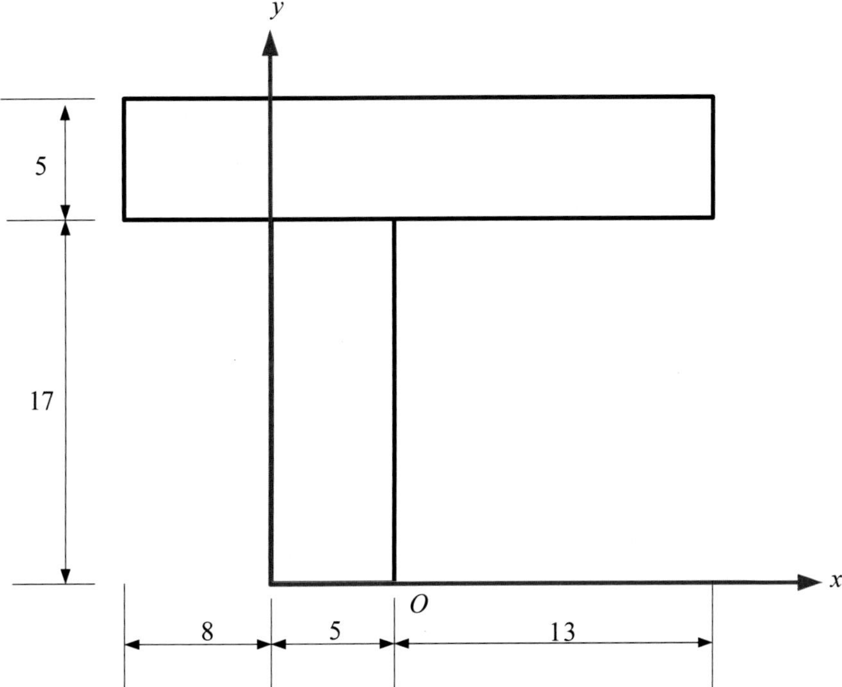

Fig. 6.7 A planar figure

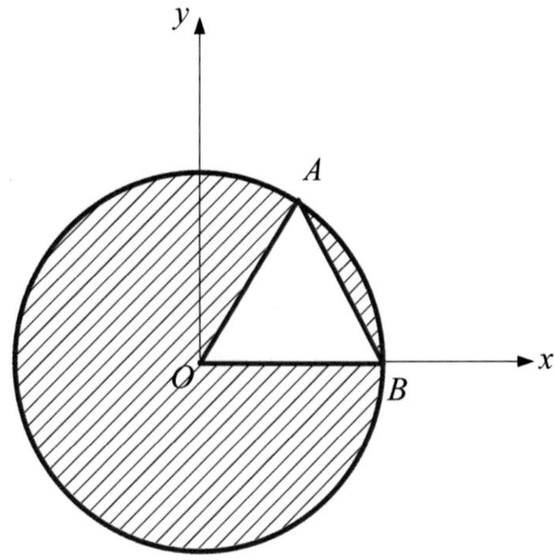

Fig. 6.8 A composite planar figure

Answers

6.1 (44.29, 0).
6.2 (4.01, 15.15).
6.3 (−0.08, −0.046).

Part II
Kinematics

In this second section, we will concentrate on the motion of a particle and the rigid body, which will pave the way to analyze the motion of machines and structures in engineering applications. In this section, we only consider the motion law in the process, and it does not deal with forces. That is to say, we only examine Kinematics from the viewpoint of geometry.

Chapter 7
Motion of a Particle

Abstract In this chapter, we will learn how to depict the motion of one particle in space. We will consider the mechanics quantities in the Cartesian coordinate system and the natural coordinate system, respectively.

Keywords Particle motion · Displacement · Velocity · Acceleration

7.1 Cartesian Coordinate System

In order to characterize the motion of a point in a space, we first construct a Cartesian coordinate system, as shown in Fig. 7.1.

We use a displacement vector or radius vector $\boldsymbol{r}$ from the origin point O, pointing to the point M to depict the position of this point. In the Cartesian coordinate system, the coordinate of the point M can be written as

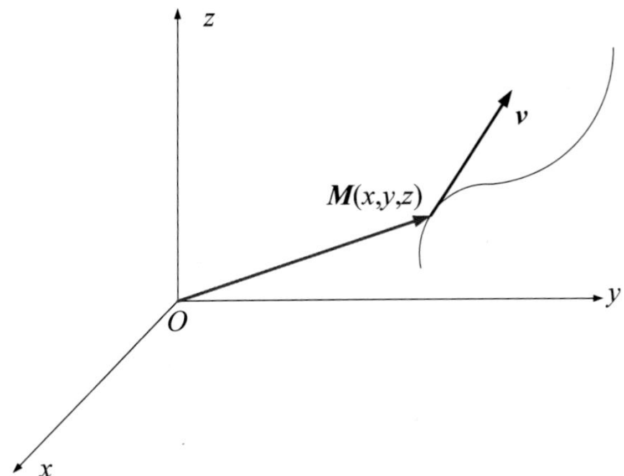

Fig. 7.1 A point in space

J. Liu, *Lecture Notes on Theoretical Mechanics*,
https://doi.org/10.1007/978-981-13-8035-8_7

$$\boldsymbol{r} = x\boldsymbol{i} + y\boldsymbol{j} + z\boldsymbol{k}.$$

Obviously, one displacement vector corresponds to one position of the point, so it can uniquely demonstrate the motion of the point. The displacement vector is actually a function of the time t, so it can be further expressed as

$$\boldsymbol{r}(t) = x(t)\boldsymbol{i} + y(t)\boldsymbol{j} + z(t)\boldsymbol{k}.$$

The velocity of point M is in the tangent direction of the orbit of the point in the space, which is shown in Fig. 7.1. The definition of the velocity is the derivative of the displacement vector

$$\boldsymbol{v} = \lim_{\Delta t \to 0} \frac{\Delta \boldsymbol{r}}{\Delta t} = \frac{\mathrm{d}\boldsymbol{r}}{\mathrm{d}t} = \dot{\boldsymbol{r}},$$

where the dot above a character denotes the derivative with respect to time. In the Cartesian coordinate system, the velocity is further written as

$$\begin{aligned} \boldsymbol{v}(t) &= \dot{\boldsymbol{r}}(t) = \dot{x}(t)\boldsymbol{i} + \dot{y}(t)\boldsymbol{j} + \dot{z}(t)\boldsymbol{k} \\ &= v_x(t)\boldsymbol{i} + v_y(t)\boldsymbol{j} + v_z(t)\boldsymbol{k}. \end{aligned}$$

Then we have the following relation:

$$\begin{cases} v_x = \dot{x} \\ v_y = \dot{y} \\ v_z = \dot{z} \end{cases}.$$

In order to depict the change rate of the velocity, the acceleration, which was introduced by Galileo, can be defined as

$$\boldsymbol{a} = \lim_{\Delta t \to 0} \frac{\Delta \boldsymbol{v}}{\Delta t} = \frac{\mathrm{d}\boldsymbol{v}}{\mathrm{d}t} = \dot{\boldsymbol{v}} = \ddot{\boldsymbol{r}},$$

and the component relations are as follows:

$$\begin{aligned} \boldsymbol{a}(t) &= \ddot{\boldsymbol{r}}(t) = \ddot{x}(t)\boldsymbol{i} + \ddot{y}(t)\boldsymbol{j} + \ddot{z}(t)\boldsymbol{k} \\ &= \dot{\boldsymbol{v}}(t) = \dot{v}_x(t)\boldsymbol{i} + \dot{v}_y(t)\boldsymbol{j} + \dot{v}_z(t)\boldsymbol{k} \\ &= a_x(t)\boldsymbol{i} + a_y(t)\boldsymbol{j} + a_z(t)\boldsymbol{k}. \end{aligned}$$

Namely, we have

$$\begin{cases} a_x = \dot{v}_x = \ddot{x} \\ a_y = \dot{v}_y = \ddot{y} \\ a_z = \dot{v}_z = \ddot{z} \end{cases}.$$

7.2 Natural Coordinate System

Normally, we use another system to investigate the motion of a particle, i.e., the natural coordinate system. In this case, we adopt the arc length and slope angle at an arbitrary point to determine the position of the point. In this situation, the physical parameters are all the function of the time t. For example, the arc length $s = s(t)$ and the slope angle $\theta = \theta(t)$ (Fig. 7.2).

$$\boldsymbol{v} = \lim_{\Delta t \to 0} \frac{\Delta \boldsymbol{r}}{\Delta t} = \lim_{\Delta t \to 0} \frac{\Delta s}{\Delta t} \boldsymbol{\tau} = \dot{s} \boldsymbol{\tau},$$

where the vector $\boldsymbol{\tau}$ is a unit vector ($|\boldsymbol{\tau}| = 1$), which is along the tangential line of the point M, and in other words, it is of the same direction of the velocity. According to the definition of the velocity, the velocity is expressed as

$$\boldsymbol{v} = v \boldsymbol{\tau}.$$

In combination with the former related equations, one has

$$v = \dot{s}.$$

Next, we examine the expression of the acceleration. According to the definition, the acceleration is

$$\boldsymbol{a} = \dot{\boldsymbol{v}} = \dot{v} \boldsymbol{\tau} + v \dot{\boldsymbol{\tau}}.$$

In the above expression, it is normally forgotten to write the second term. In fact, the unit vector $\boldsymbol{\tau}$ is also a function with respect to the time t. It can be proven (but

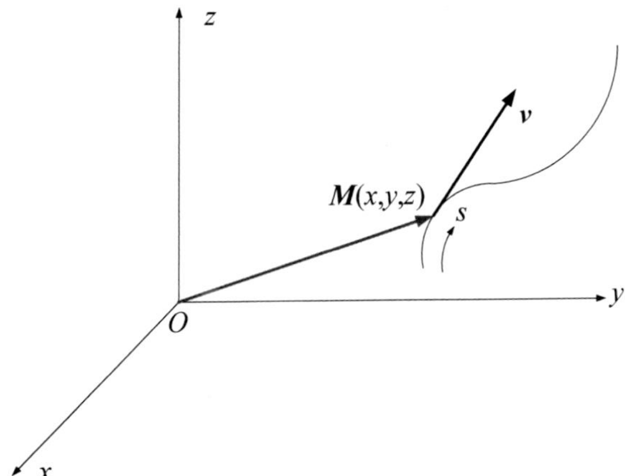

Fig. 7.2 Natural coordinate system

here we ignore the derivation) that the direction of $\dot{\boldsymbol{\tau}}$ is normal to the unit vector $\boldsymbol{\tau}$. We then define a new normal vector $\boldsymbol{n}$, and $\boldsymbol{\tau} \cdot \boldsymbol{n} = 0$. It can also be proven that

$$\dot{\boldsymbol{\tau}} = \frac{v}{\rho}\boldsymbol{n},$$

where ρ is the curvature radius of the curve at one point. Accordingly, the acceleration is

$$\boldsymbol{a} = \dot{\boldsymbol{v}} = \dot{v}\boldsymbol{\tau} + \frac{v^2}{\rho}\boldsymbol{n}.$$

This means that the acceleration includes two portions, where one is along the $\boldsymbol{n}$ direction and the other is along $\boldsymbol{\tau}$ direction, so it can be expressed as

$$\boldsymbol{a} = \boldsymbol{a}_\tau + \boldsymbol{a}_n = a_\tau \boldsymbol{\tau} + a_n \boldsymbol{n}.$$

In comparison with the former related equations, one has

$$a_\tau = \dot{v},$$

$$a_n = \frac{v^2}{\rho}.$$

The above two accelerations are, respectively, named as the tangential acceleration and the centripetal or normal acceleration. The full acceleration is the resultant vector of these two accelerations, and they satisfy the parallelogram law, which is schematized in Fig. 7.3. The magnitude of the full acceleration is

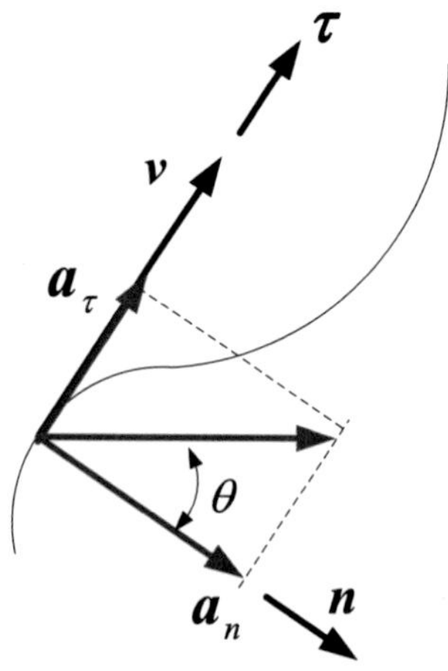

Fig. 7.3 Acceleration

$$a = \sqrt{a_\tau^2 + a_n^2},$$

and the angle is expressed as

$$\tan\theta = \frac{|a_\tau|}{a_n}.$$

Summary

In this chapter, we must master how to depict the motion of one particle in space, including such parameters as displacement, velocity, and acceleration. We consider their mechanics quantities in the Cartesian coordinate system and the natural coordinate system, respectively.

Exercises

7.1 As shown in Fig. 7.4, a particle is in a circular motion, from the origin O to an arbitrary point A. The radius of the circle is $R = 2$ mm, and the angular displacement $\varphi = 3t - \frac{1}{2}t^2$. When $t = 1$ s, please calculate the values of the angular velocity ω, the angular acceleration ε, the arc length s, the velocity, and acceleration of point A. Please also schematize the velocity and acceleration of point A in the fiugre.

7.2 Please derive the expressions of the tangential and normal accelerations in the natural coordinate system.

7.3 As shown in Fig. 7.5, a bar AB moves upward with the uniform velocity $\boldsymbol{u}$, and at the initial state, the angle $\phi = 0$. The bar OC rotates with respect to point O, and there is a jacket linking the bar AB at point A. When the angle $\phi = 30°$, please calculate the velocity and acceleration for point C. The angular velocity and angular acceleration of the bar are denoted by ω and ε, respectively.

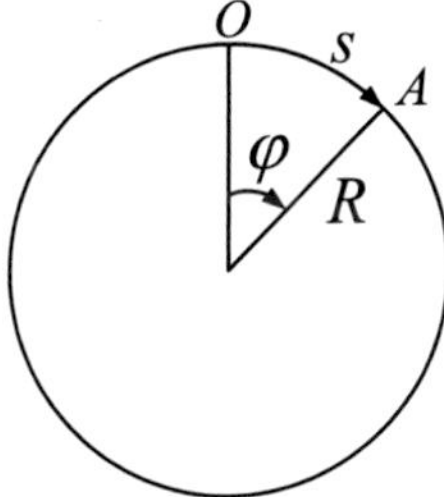

Fig. 7.4 A particle in a circular motion

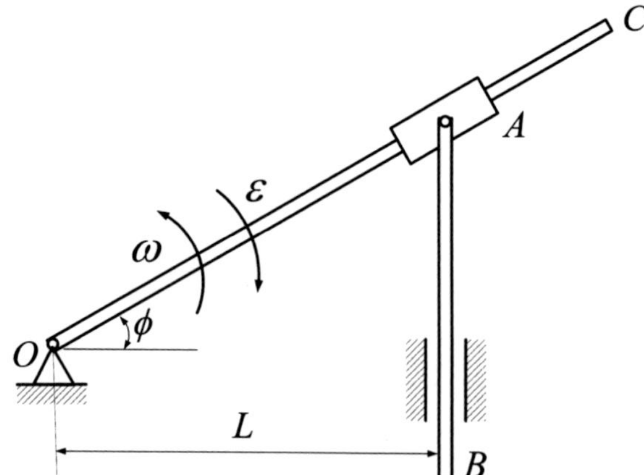

Fig. 7.5 A mechanical system with two bars

Answers

7.1 $\omega = 3 - t\,\text{rad/s},\ \varepsilon = -1\,\text{rad/s}.$

7.3 $v_c = \omega L_{OC},\, a_c^n = \omega^2 L_{OC},\, a_c^\tau = \varepsilon L_{OC}.$

Chapter 8
Basic Motion of the Rigid Body

Abstract In this chapter, we will learn two basic motions for the rigid body. The first one is translation and the second one is the rotation.

Keywords Basic motion · Translation · Rotation to an axis

We have just learned the motion of a particle in space, but throughout this chapter, we mainly concentrate on the motion of the rigid body, which is more useful and complex in practice. In this chapter, we introduce the basic motion of the rigid body, i.e., the translation and rotation with a fixed axis.

8.1 Translation

Translation is the motion of a rigid body, not a particle, as we all know the fact that the rigid body is the cluster system of particles. As shown in Fig. 8.1, a rigid body is lying on a desk undergoing a horizontal force. This object can move linearly in the desk surface, so clearly, its motion is called translation. However, what's the actual definition of translation? Normally, we can draw a straight line on the object in the initial state, if after a while the current position of the line is always parallel to the initial line, then the motion of the object is defined as translation.

Let's look at another example shown in Fig. 8.2. A rigid body is tied by two slender ropes, which are hung on the ceiling. The two parallel bars can rotate on the ceiling, and the rigid body can sweep with the rotation of the two ropes. We examine the motion law of one point *A* on the rigid body, whose orbit is really a portion of a circle. Another point *B* is also in a circular motion. However, in this case, the motion

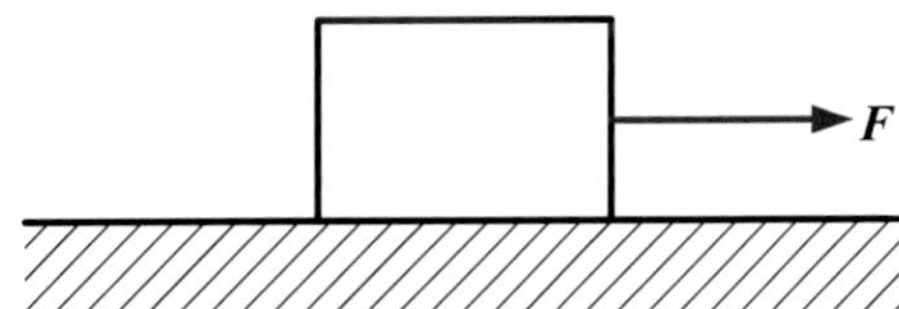

Fig. 8.1 Translation of an object

J. Liu, *Lecture Notes on Theoretical Mechanics*,
https://doi.org/10.1007/978-981-13-8035-8_8

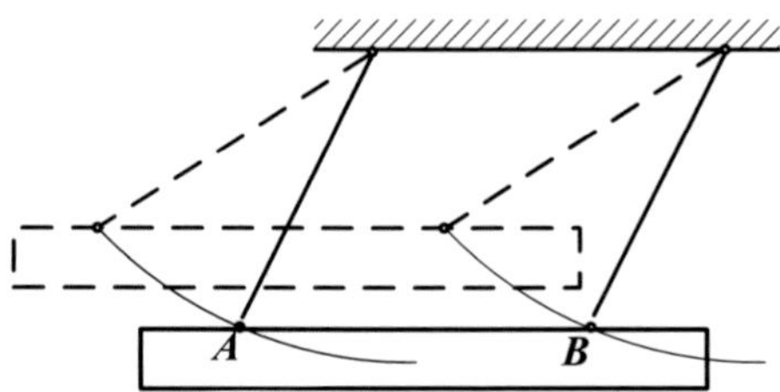

Fig. 8.2 Translation of a bar

type of the rigid body in Fig. 8.2 is translation. We also analyze a straight line, such as AB. After the rigid body moves to another position, such as the virtual line in Fig. 8.2, the current straight line is parallel to AB. Therefore, although an arbitrary point on the rigid body is in a circular motion, the motion of the total system is surely translation.

It's natural to see that the velocities and accelerations of point A and B are both equal, as the length of AB on the rigid body does never change. As a result, we arrive at the conclusion that every point on a rigid body in translation has the same velocity and acceleration. For two arbitrary points, one has

$$\boldsymbol{v}_A = \boldsymbol{v}_B,$$
$$\boldsymbol{a}_A = \boldsymbol{a}_B.$$

8.2 Rotation on a Fixed Axis

As shown in Fig. 8.3, if we impose a moment on a door, the door will be closed with respect to a fixed axis. This type of motion is also a basic motion, termed as rotation with respect to a fixed axis. In this rotation process, there always exists a fixed axis on the rigid body. This type of motion takes some special features. If we look through the positive direction of z-axis in Fig. 8.3, we can find the orbit of an arbitrary point on the rigid body in rotation, which is actually a circle.

To depict the motion of the rigid body, we introduce the angular displacement φ, whose sign also obeys the right-hand screw law. The angular velocity can be defined as

$$\omega = \lim_{\Delta t \to 0} \frac{\Delta\varphi}{\Delta t} = \dot{\varphi}.$$

The angular velocity can also be related with the frequency via

$$\omega = 2\pi f.$$

Introducing the concept of rotation velocity n with the unit of 1/min, one has

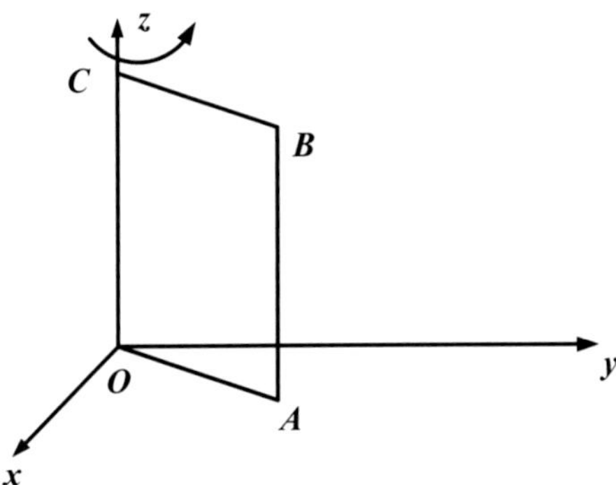

Fig. 8.3 A door in rotation

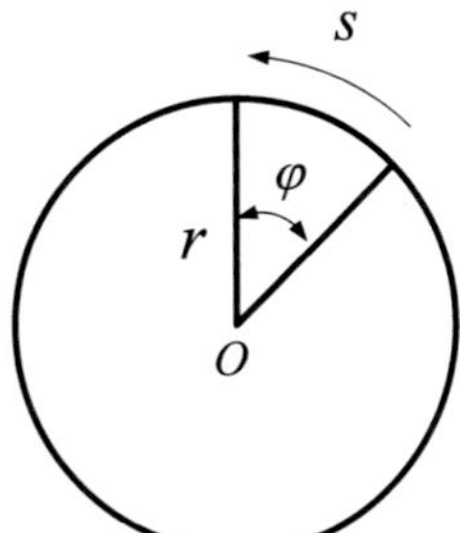

Fig. 8.4 Circular motion

$$\omega = \frac{2\pi n}{60} = \frac{\pi n}{30}.$$

Similarly, the angular acceleration can be further defined as

$$\varepsilon = \lim_{\Delta t \to 0} \frac{\Delta \omega}{\Delta t} = \dot{\omega} = \ddot{\varphi}.$$

The above relation can be analogous to the displacement, velocity, and acceleration defined in the last chapter.

If the rigid body is in rotation with respect to an axis, any point on the rigid body is in a circular motion. As shown in Fig. 8.4, the radius r is the vertical distance from the axis to the arbitrary point, s is the arc length, and φ is the corresponding angular displacement. We then have the following geometric relation:

$$s = r\varphi.$$

Taking derivatives on both sides of the above equation, one has

$$v = \dot{s} = r\dot{\varphi} = \omega r,$$
$$a_\tau = \dot{v} = \ddot{s} = r\ddot{\varphi} = \varepsilon r.$$

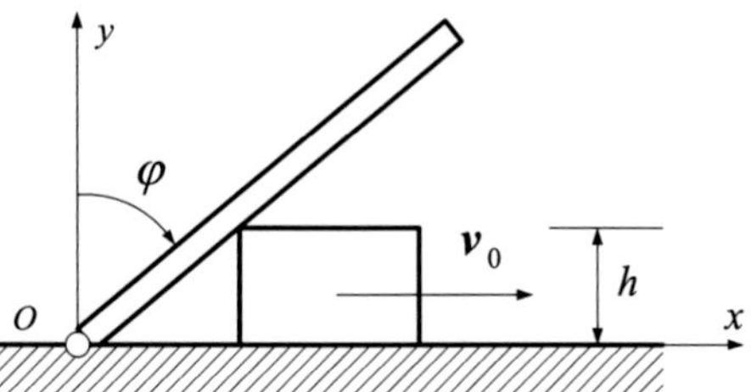

Fig. 8.5 A bar and an object

Moreover, we have the following relations:

$$a_n = \frac{v^2}{r} = \omega^2 r = \omega v.$$

Example 1 As shown in Fig. 8.5, there is a slender bar fixed at one point, and it is supported by an object moving at a constant velocity $\boldsymbol{v}_0$. Please write out the angular velocity and angular acceleration of the slender bar.

Answer: In the x-direction, the object has a displacement as

$$x = v_0 t.$$

We have the following geometric condition:

$$\tan\varphi = \frac{x}{h} = \frac{v_0 t}{h},$$

and then

$$\varphi = \arctan\frac{v_0 t}{h}.$$

The angular velocity is

$$\omega = \dot{\varphi} = \frac{h v_0}{h^2 + v_0^2 t^2}.$$

The angular acceleration is

$$\varepsilon = \dot{\omega} = \ddot{\varphi} = -\frac{2 h v_0^3 t}{\left(h^2 + v_0^2 t^2\right)^2}.$$

Summary

In this chapter, we should master the knowledge on two basic motions for the rigid body. The first one is translation and the second one is the rotation.

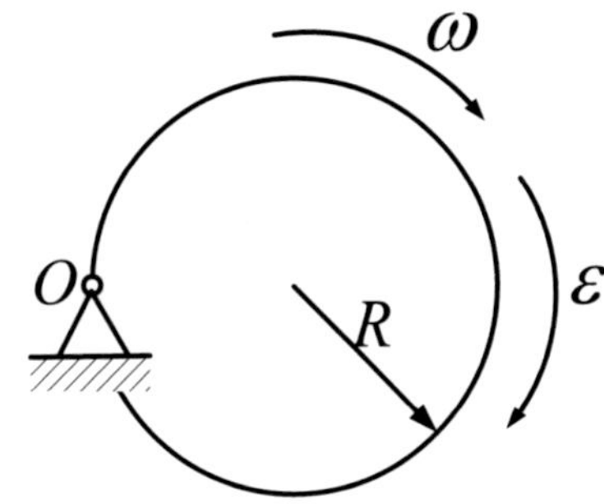

Fig. 8.6 A disk in rotation with respect to one axis

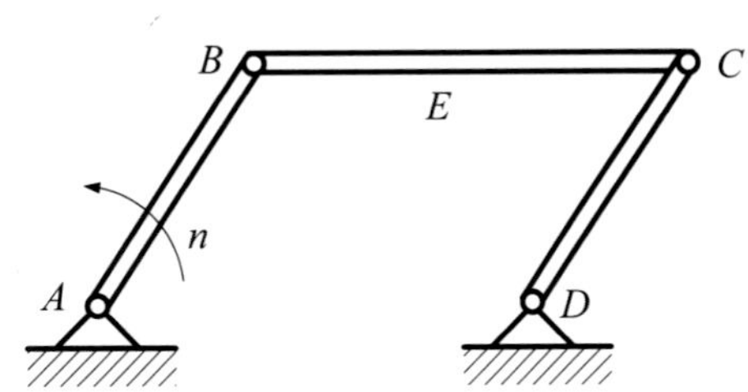

Fig. 8.7 A mechanical system

Exercises

8.1 As shown in Fig. 8.6, a circular disk is rotating with respect to the axis O. Its weight is W and its radius is R. At one certain time, the angular velocity of the disk is ω and angular acceleration is ε. Please give the expressions of the velocity and acceleration on the center of the disk.

8.2 What is the difference between one particle and one rigid body in translation? Please give the figure comparison.

8.3 As shown in Fig. 8.7, there are three bars moving. The bar AB is parallel to CD at any time, whose rotation velocity is $n = 300$ r/min. The length of the bar $AB = CD = 150$ mm. Please give the velocity and acceleration of the middle point of BC, i.e., point E.

Answers

8.1 $v_c = \omega R, a_c^n = \omega^2 R, a_c^\tau = \varepsilon R.$

8.3 $v_E = 1.5\pi\,\text{m/s}, a_E^n = 0\,\text{m/s}^2, a_E^\tau = 15\pi\,\text{m/s}^2.$

Chapter 9
Planar Motion of the Rigid Body

Abstract In this chapter, we will learn the fundamental knowledge on the planar motion for a rigid body. First, the concept of relative velocity is introduced, and one can get the composite velocity. Then we introduce the concept on the instantaneous center for the planar motion, which is convenient to analyze the velocity of each point on the rigid body.

Keywords Planar motion · Relative velocity · Instantaneous center

Besides translation and rotation, planar motion is another common motion in *Theoretical Mechanics*. If an arbitrary point on the rigid body has a constant distance with a fixed plane, then this motion is called the planar motion. We will focus on the motion of a planar figure in this chapter, as it can represent the motion of the rigid body.

9.1 Relative Velocity

We consider two points on the planar figure A and B, which have the velocity $\boldsymbol{v}_A$ and $\boldsymbol{v}_B$, respectively, as schematized in Fig. 9.1. We normally name point A as the base point, as it is a reference point. There is a relative velocity $\boldsymbol{v}_{BA}$, with the meaning that A is the base point. As a result, we know that $\boldsymbol{v}_{BA}$ is not equal to $\boldsymbol{v}_{AB}$. According to the velocity superposition, one has

$$\boldsymbol{v}_B = \boldsymbol{v}_A + \boldsymbol{v}_{BA},$$

where the direction of the relative velocity $\boldsymbol{v}_{BA}$ is perpendicular to the line AB, as point B rotates with respect to point A in a circular motion. From the above formula, we can solve the velocity of any point on the planar figure if the base point is given. Therefore, this method of velocity composition is called "Method of base point".

As a consequence, if we decompose the above velocities in the direction of line AB, then one has

J. Liu, *Lecture Notes on Theoretical Mechanics*,
https://doi.org/10.1007/978-981-13-8035-8_9

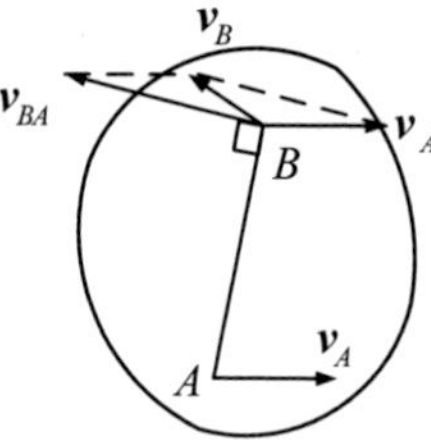

Fig. 9.1 A rigid body in planar motion

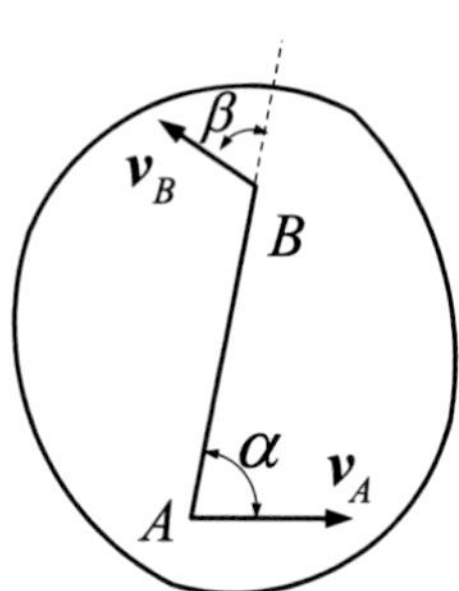

Fig. 9.2 Velocity projection

$$\begin{aligned} \boldsymbol{v}_B|_{AB} &= \boldsymbol{v}_A|_{AB} + \boldsymbol{v}_{BA}|_{AB} \\ &= \boldsymbol{v}_A|_{AB}. \end{aligned}$$

This means the projections of the velocities of the two points are equal along their connection line. In fact, this is the second method to study the velocity composition, which is termed as "Method of velocity projection". As shown in Fig. 9.2, if we have known the angles between the velocities and line *AB*, we then have the relation

$$v_A \cos\alpha = v_B \cos\beta.$$

9.2 Instantaneous Center

Next, we will introduce a simple and convenient method, i.e., the instantaneous center method. It is true that there is always one special point on the planar figure with zero velocity. Herein we omit the process of proof, and we only remember this conclusion. This point is called "instantaneous center of velocity", or "instantaneous center". If P is the instantaneous center, then $v_P = 0$. We also adopt the method of base point, where we select the instantaneous center as the base point. Consequently, we have

$$\begin{aligned} \boldsymbol{v}_A &= \boldsymbol{v}_P + \boldsymbol{v}_{AP} \\ &= \boldsymbol{v}_{AP}. \end{aligned}$$

From the above discussions, we know that the velocity direction of point A is normal to the line AP. It should be noticed that P is only a point with an **instantaneous** zero velocity, and we do not know what the state in the next time will be. At this instantaneous time, point P is similar to a circle center, and point A is in a circular motion only at this time. If the angular velocity of the rigid body is ω, the velocity of point A is

$$v_A = v_{AP} = \omega \cdot AP.$$

In contrast, how to find the instantaneous center if the velocity is given? Let us look at a planar figure with two velocities of two points, as shown in Fig. 9.3.

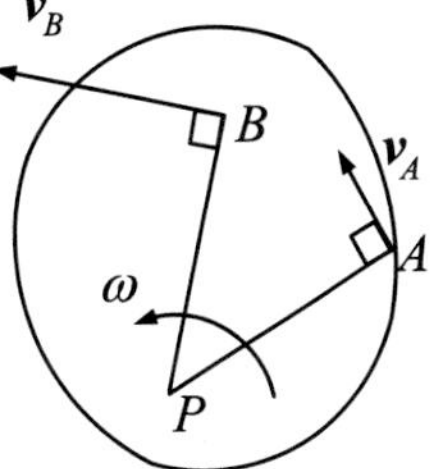

Fig. 9.3 Instantaneous center

If the velocities of points A and B are already known, then one has $\boldsymbol{v}_A \perp AP$, $\boldsymbol{v}_B \perp BP$. Therefore, the instantaneous center P must be the cross point of line BP and AP. However, in this case, there is a condition that $\boldsymbol{v}_A$ is not parallel to $\boldsymbol{v}_B$. How about the case that they are parallel?

Let us look at Figs. 9.4 and 9.5, where the velocities of the two points are parallel to each other.

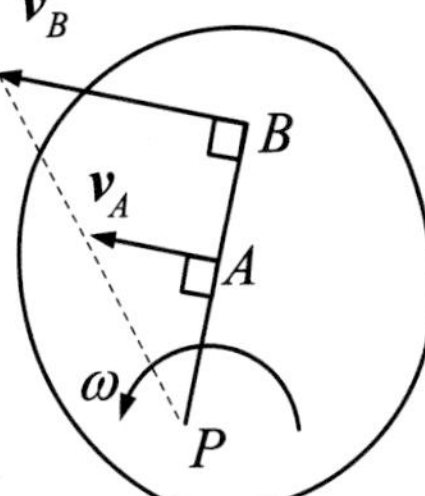

Fig. 9.4 Instantaneous center

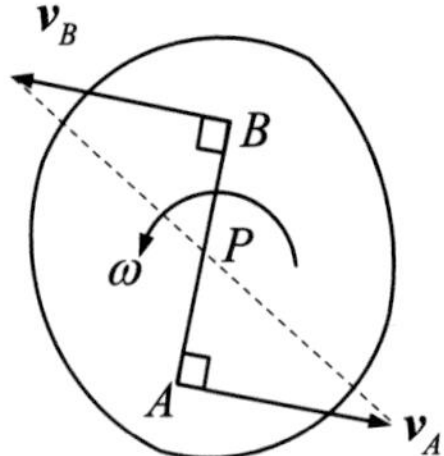

Fig. 9.5 Instantaneous center

As we know that

$$v_A = \omega \cdot AP,$$
$$v_B = \omega \cdot BP,$$

one has the relation

$$\frac{v_A}{v_B} = \frac{AP}{BP}.$$

The other conditions are that $\boldsymbol{v}_A \perp AP$ and $\boldsymbol{v}_B \perp AP$. Therefore, the instantaneous center can be determined in Fig. 9.4. A similar result can also be schematized in Fig. 9.5. In this case, the two velocities are parallel, but they are in the opposite directions. The magnitudes of the two velocities are not equal, so one can determine the instantaneous center according to the proportional relation.

Two special cases are shown in Figs. 9.6 and 9.7, where the two velocities with the same magnitudes are parallel, and in the same direction.

In this scenario, we can't find out where the instantaneous center is, and we call this state "instantaneous translation". At this time, the angular velocity is zero.

The last case is a wheel purely rolling on a substrate, as shown in Fig. 9.8. Pure rolling means no sliding on the interface. This is because that, at the contact point, the two points at the roller and the substrate have the same velocity. However, the substrate is static, so the velocity of the contact point on the roller is zero. Namely, it is right the instantaneous center.

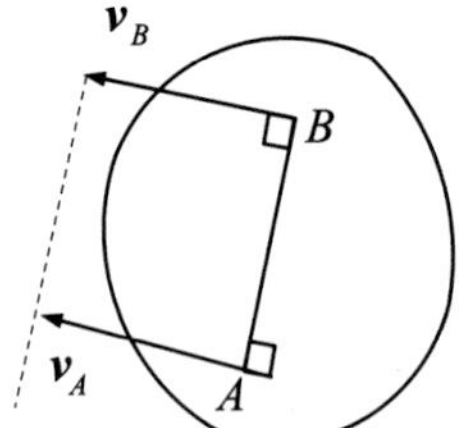

Fig. 9.6 Instantaneous translation

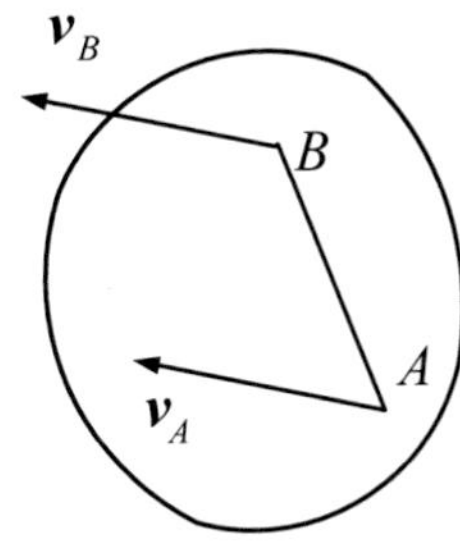

Fig. 9.7 Instantaneous translation

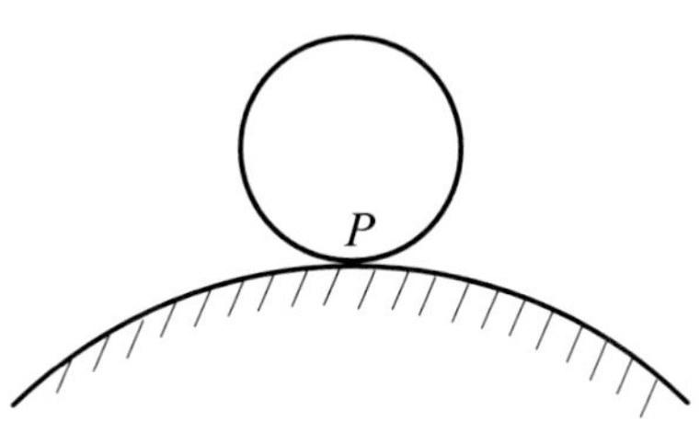

Fig. 9.8 Instantaneous center

Example 1 As shown in Fig. 9.9, a slender bar is sliding at a wall corner, with the length L, angle θ, angular velocity ω (unknown), $\boldsymbol{v}_A = \boldsymbol{u}$. Please get the expressions of ω and $\boldsymbol{v}_B$.

The motion type of the slender bar is in planar motion. The velocity directions of points A and B can be determined, as shown in Fig. 9.9. Then the position of the instantaneous center can be found, as shown in Fig. 9.9. One has the relation

$$v_A = \omega \cdot AP = \omega L \sin\theta = u,$$
$$v_B = \omega \cdot BP = \omega L \cos\theta.$$

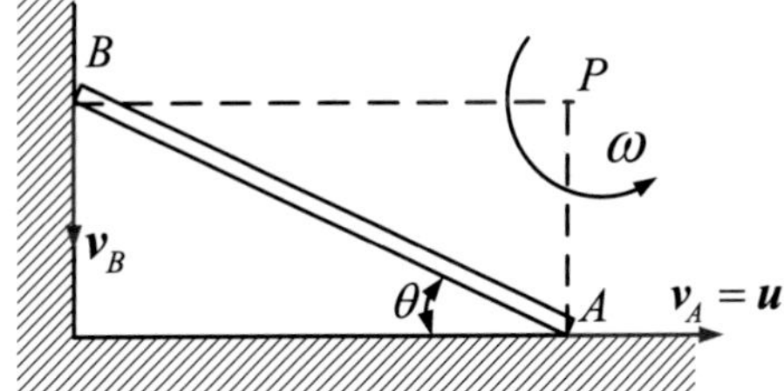

Fig. 9.9 Planar motion

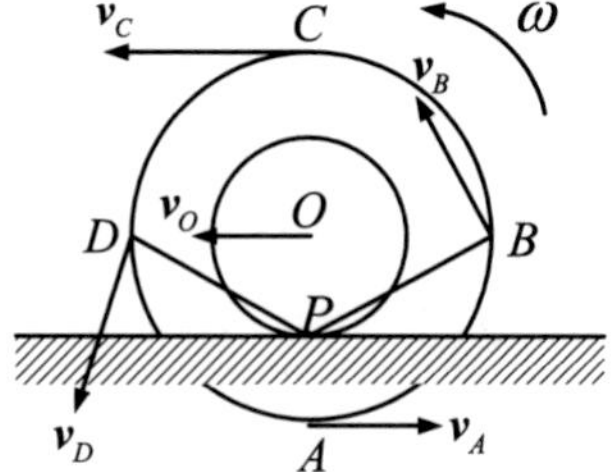

Fig. 9.10 A wheel in planar motion

Then we have

$$\omega = \frac{u}{L \sin\theta},$$
$$v_B = \omega L \cos\theta = u \cot\theta.$$

Example 2 As shown in Fig. 9.10, a train wheel is in pure rolling. The outer radius is R, and the inner radius is r. The velocity of the wheel center is $\boldsymbol{v}_O = \boldsymbol{v}$. Please give the velocity expressions of point A, B, C, and D.

It is judged that the wheel is in planar motion. Due to pure rolling, point P is the instantaneous center:

$$v_A = \omega \cdot AP = \omega(R - r),$$
$$v_B = \omega \cdot BP = \omega\sqrt{R^2 + r^2} = v_D,$$
$$v_O = \omega \cdot OP = \omega r = v,$$
$$v_C = \omega \cdot CP = \omega(R + r).$$

Then one has

$$\omega = \frac{v}{r},$$
$$v_A = \omega(R - r) = \frac{R - r}{r} v,$$
$$v_B = v_D = \omega\sqrt{R^2 + r^2} = \frac{\sqrt{R^2 + r^2}}{r} v,$$
$$v_C = \omega(R + r) = \frac{R + r}{r} v.$$

Summary

In this chapter, we need to master the knowledge on the planar motion for a rigid body. We should know the concept of relative velocity and composite velocity, and the instantaneous center for the rigid body; then the velocity analysis for each point on the rigid body can be made.

Exercises

9.1 As shown in Fig. 9.11, a square with the side length L is rotating with respect to point O, and the angular velocity and angular acceleration are ω and ε, respectively. Write down the expressions of the velocity and acceleration for point A, and then depict them in the figure.

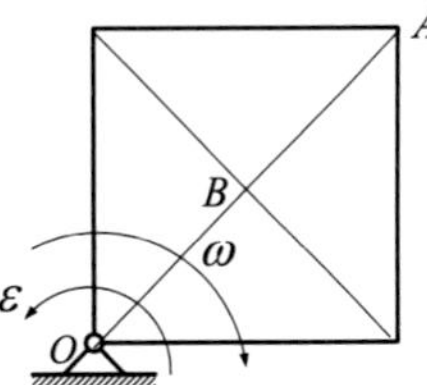

Fig. 9.11 A square rotating with respect to point O

9.2 As shown in Fig. 9.12, a bar AB is sliding downward at a wall corner and point C is the mass center of the bar. The angle $\varphi = 45°$, the velocity of point A is $\boldsymbol{v}_A = 5\boldsymbol{u}$, the length of the bar is L, and its mass is m. Please give the values of the velocities of point B $\boldsymbol{v}_B$ and point C $\boldsymbol{v}_C$, angular velocity of the bar ω of the bar.

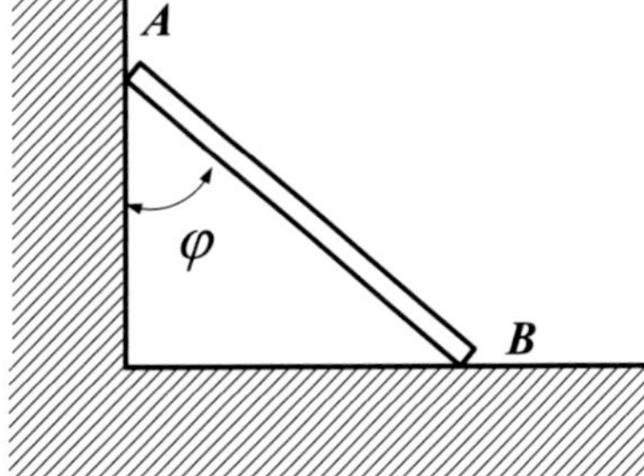

Fig. 9.12 One bar sliding on the surfaces

9.3 As shown in Fig. 9.13, there is a uniformly circular disk which is fixed with a curved bar ABD. The radius of the disk is R, whose center is C. The angle $B = 90°$, and $AB = BD = R$. The disk is in pure rolling on the ground, and the angular velocity of the disk is ω. Both of the disk and bar are of the same mass

m. Please give the velocities of point A, B, C, and D.

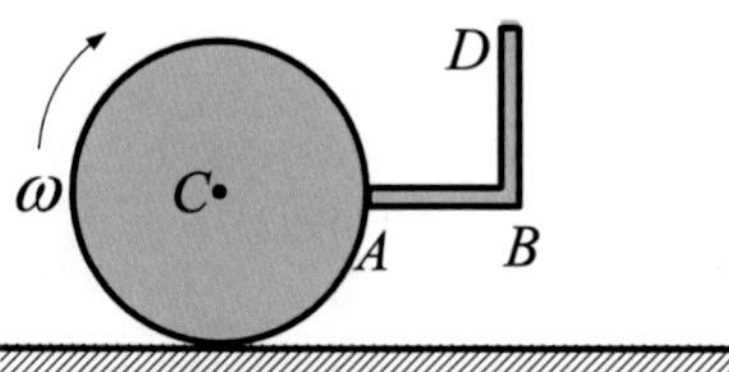

Fig. 9.13 A disk fixed with a bar in pure rolling

9.4 As shown in Fig. 9.14, a wheel B with the radius r is in pure rolling on a fixed wheel O with the radius R, and $R = 2r$. The bar A links the slider A and the wheel B, and the slider is sliding in the vertical trough. The length of AB is $l = 2\sqrt{2}r$. At one certain time, point B and point O are in the same vertical line, and $\alpha = 45°$. The velocity of point B is $\boldsymbol{v}_B$, and the angular acceleration is ε_B. Please calculate the velocity and acceleration of the slider A.

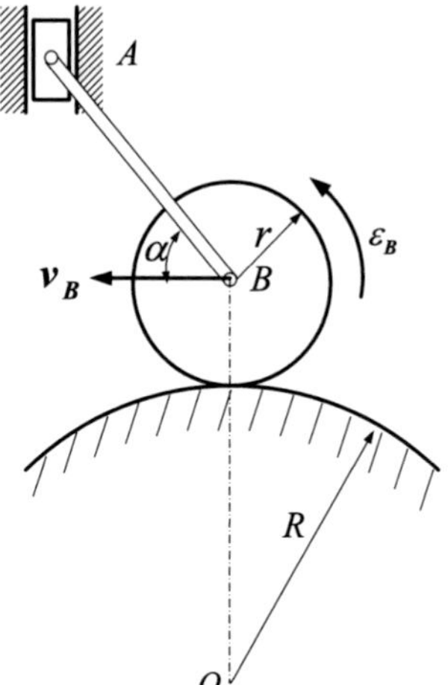

Fig. 9.14 A complex mechanical system

9.5 As shown in Fig. 9.15, a disk is linked with a bar at the wall corner. The disk is in pure rolling on the ground, with the radius $r = 1$ m. The length of the bar is $AB = 3$ m, and the angle between AB and the vertical direction is 60°. The angular velocity of the disk is $\omega = 1$ rad/s, and the angular acceleration is $\varepsilon = 1$ rad/s^2. Please solve the velocity and acceleration of point B at the bar.

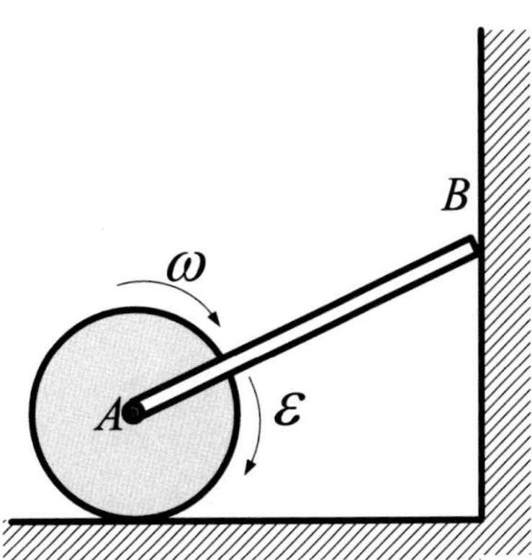

Fig. 9.15 A disk linked with a bar at the wall corner

Answers

9.1 $v_A = \sqrt{2}\omega L$, $a_A^\tau = \sqrt{2}\varepsilon L$, $a_A^n = \sqrt{2}\omega^2 L$
9.2 $v_B = 5u$, $v_C = \frac{5\sqrt{2}}{2}u$, $\omega = \frac{5\sqrt{2}u}{L}$
9.3 $v_A = \sqrt{2}\omega R$, $v_B = \sqrt{5}\omega R$, $v_C = \omega R$, $v_D = 2\sqrt{2}\omega R$
9.4 $v_A = v_B$, $\varepsilon_A = \varepsilon_B$
9.5 $v_B = \sqrt{3}\,\mathrm{m/s}$, $\varepsilon_B = \sqrt{3}\,\mathrm{m/s^2}$.

Part III
Kinetics

In this section, we will learn the relations between motion and force, which are more complex and interesting. This section is aimed to solve some practical problems in astronomy and mechanical motions. We will learn the concepts of Newton's second law, momentum, moment of momentum, and kinetic energy.

Chapter 10
Newton's Laws of Motion

Abstract In this chapter, we will learn the Newton's laws of motion, especially on the second law, where the motion equation for one particle can be given.

Keywords Newton's law of motion · Newton's second law · Differential equation for a particle in motion

The great scientist in human history, Newton developed calculus and derived the law of universal gravitation. Besides, based upon the theories of some giants, he proposed Newton's laws of motion. The first law is the law of inertia, which declares that "No effect of external forces, the object will continue to be in the uniform linear motion." The third law is the law of action and reaction, which is one of the fundamental axioms in Statics. Herein, we mainly concentrate on the Newton's second law, which can be expressed as

$$\boldsymbol{F} = m\boldsymbol{a},$$

where $\boldsymbol{F}$ is the resultant external force of the object, and $\boldsymbol{a}$ is the acceleration of the object. This is a vector equation, which is not easy to solve. Therefore, we can investigate it in the Cartesian coordinate system. It can be further written as

$$\begin{cases} m\ddot{x} = \Sigma X \\ m\ddot{y} = \Sigma Y \\ m\ddot{z} = \Sigma Z. \end{cases}$$

According to the above formula of Newton's second law, we have two types of problems to solve. One is that if the motion is given, how to solve the force; and the other is that if the force is given, how to solve the motion. The first type is easier, for it only needs calculus operations. However, the second type needs to solve the ordinary differential equations, where the initial conditions are very important.

J. Liu, *Lecture Notes on Theoretical Mechanics*,
https://doi.org/10.1007/978-981-13-8035-8_10

Example 1 The motion of a planet is given as

$$\begin{cases} x = A\cos\omega t \\ y = B\sin\omega t \end{cases},$$

then we can get the velocity as

$$\begin{cases} \dot{x} = -\omega A \sin\omega t \\ \dot{y} = \omega B \cos\omega t \end{cases},$$

and the acceleration is

$$\begin{cases} \ddot{x} = -\omega^2 A \cos\omega t = a_x \\ \ddot{y} = -\omega^2 B \sin\omega t = a_y. \end{cases}$$

According to Newton's second law, one has

$$\begin{cases} \Sigma X = -m\omega^2 A \cos\omega t \\ \Sigma Y = -m\omega^2 B \sin\omega t. \end{cases}$$

Therefore, the resultant force is

$$\begin{aligned} \boldsymbol{F} &= -m\omega^2 A \cos\omega t \boldsymbol{i} - m\omega^2 B \sin\omega t \boldsymbol{j} \\ &= -m\omega^2 \boldsymbol{r}. \end{aligned}$$

The result shows that the acceleration has the opposite direction to that of the resultant force.

Example 2 An object moves directly in the air from the original position in static. The resistance force $F = \alpha v$, where v is the velocity magnitude of the object. Please determine the orbit of the object.

Answer: We establish a one-dimensional Cartesian coordinate system, as shown in Fig. 10.1.

The initial conditions are

$$t = 0, x_0 = 0, v_0 = 0.$$

The Newton's second law reads

$$m\frac{\mathrm{d}v}{\mathrm{d}t} = mg - \alpha v,$$

i.e.,

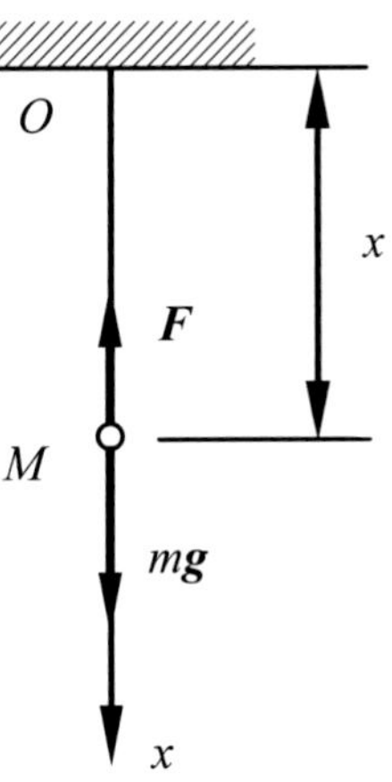

Fig. 10.1 An object dropping in the air

$$\frac{dv}{dt} = g - \mu v,$$

where $\mu = \frac{mg}{\alpha}$.

The above integration leads to

$$\int_{v_0}^{v} \frac{dv}{g - \mu v} = \int_{0}^{t} dt$$

Then the result is further given as

$$v = \mu\left(1 - e^{-\frac{g}{\mu}t}\right)$$

The further integration is

$$\int_{0}^{x} dx = \mu \int_{0}^{t} \left(1 - e^{-\frac{g}{\mu}t}\right) dt$$

That is

$$x = \mu t - \frac{\mu^2}{g}\left(1 - e^{-\frac{g}{\mu}t}\right)$$

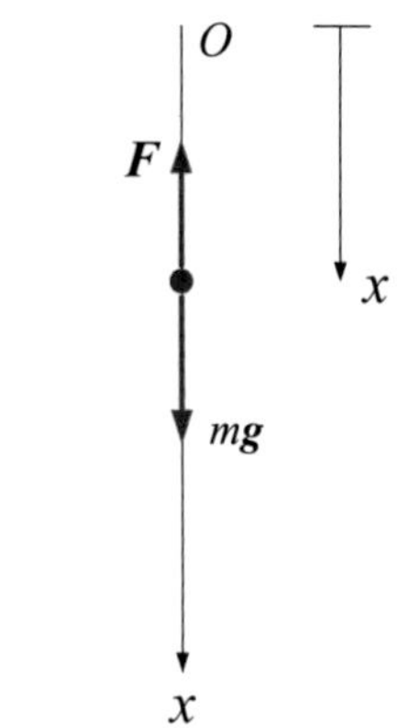

Fig. 10.2 An object falling in the liquid

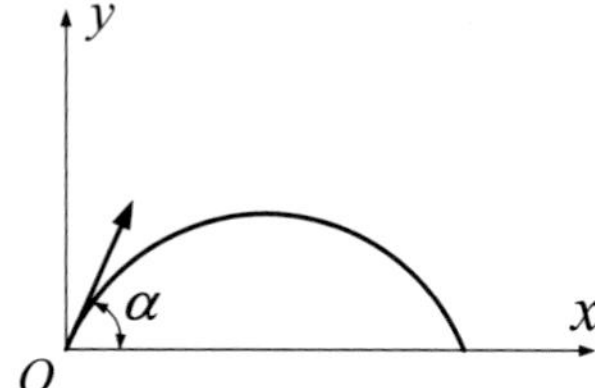

Fig. 10.3 An oblique throwing object

Summary

In this chapter, we need to master the Newton's laws of motion, especially on the second law, where the motion equation for one particle can be given. How to solve the equation also needs to be learned.

Exercises

10.1 As shown in Fig. 10.2, a small object with mass m is falling from point O in the liquid, and the resistant force of the liquid is $F = Rv^2$, where R is the resistant force factor, and v is the velocity of the object. Please give the governing equation and the solution of this problem.

10.2 As shown in Fig. 10.3, an object is in oblique throwing. The initial velocity is $\boldsymbol{v}_0$, and the initial angle is α. The resistant force is ignored, and the gravity of the object is $\boldsymbol{G}$. Please determine the orbit of this object.

Answers

10.1 $x = \frac{m}{2R} \ln \frac{mg - Rv^2}{mg}$

10.2 $y = x \tan\alpha - \frac{g}{2v_0^2 \cos^2\alpha} x^2$

Chapter 11
Theorem of Momentum

Abstract In this chapter, we will learn the theorem of momentum for one particle, and for one rigid body system. In particular, the motion theorem of the mass center for the rigid body is given.

Keywords Theorem of momentum · Conservation of the mass center · Motion theorem of the mass center

For a particle, the momentum can be defined as

$$\boldsymbol{P} = m\boldsymbol{v},$$

where m is the mass and $\boldsymbol{v}$ the velocity of the particle. The impulse in a time interval t can be given as

$$\boldsymbol{I} = \int_0^t \boldsymbol{F}\mathrm{d}t.$$

Based upon these definitions, the theorem of momentum of the particle can be derived from Newton's second law:

$$\begin{aligned} \boldsymbol{F} = m\boldsymbol{a} &= m\frac{\mathrm{d}\boldsymbol{v}}{\mathrm{d}t} \\ &= \frac{\mathrm{d}(m\boldsymbol{v})}{\mathrm{d}t} \\ &= \frac{\mathrm{d}\boldsymbol{P}}{\mathrm{d}t}. \end{aligned}$$

For a system with a lot of particles, the momentum is the summation of the momentums of all the particles

$$\boldsymbol{P} = \Sigma m_i \boldsymbol{v}_i.$$

J. Liu, *Lecture Notes on Theoretical Mechanics*,
https://doi.org/10.1007/978-981-13-8035-8_11

Recalling the former equation, i.e., the theorem of the resultant moment

$$m\mathbf{r}_C = \Sigma m_i \boldsymbol{r}_i,$$

we have

$$m\mathbf{v}_C = \Sigma m_i \boldsymbol{v}_i.$$

Consequently, the momentum for the particle system is

$$\boldsymbol{P} = \Sigma m_i \boldsymbol{v}_i = m\boldsymbol{v}_C$$

where $\boldsymbol{v}_C$ is the velocity of the mass center for the system or the rigid body.

For the particle system or the rigid body, the theorem of momentum can be expressed as

$$\frac{\mathrm{d}\boldsymbol{P}}{\mathrm{d}t} = \Sigma \boldsymbol{F}_i.$$

The integration to the time t yields

$$\boldsymbol{P}_2 - \boldsymbol{P}_1 = \Sigma \boldsymbol{I}_i = \Sigma \boldsymbol{F}_i.$$

The notion $\Sigma \boldsymbol{F}_i$. means the total external forces acting on the rigid body or the particle system. The theorem can be further derived as

$$\Sigma \boldsymbol{F}_i = \frac{\mathrm{d}\boldsymbol{P}}{\mathrm{d}t} = \frac{\mathrm{d}(m\boldsymbol{v}_C)}{\mathrm{d}t} = ma_C.$$

However, we often use the more convenient format of the above equation in the Cartesian coordinate system:

$$\begin{cases} ma_{Cx} = \Sigma X \\ ma_{Cy} = \Sigma Y. \end{cases}$$

This is the motion theorem of the mass center for the rigid body. If the external force is zero, the acceleration of the mass center is zero. It is noted that the motion of the mass center is conservative in the absence of the resultant force.

Example 1 A slender bar with the mass center C is sliding from the static state downward the substrate, as shown in Fig. 11.1. The substrate surface is smooth. The bar is subjected to the gravity and the vertical reaction force of the substrate.

Answer: According to the motion theorem of the mass center, one has

$$ma_{Cx} = \Sigma X.$$

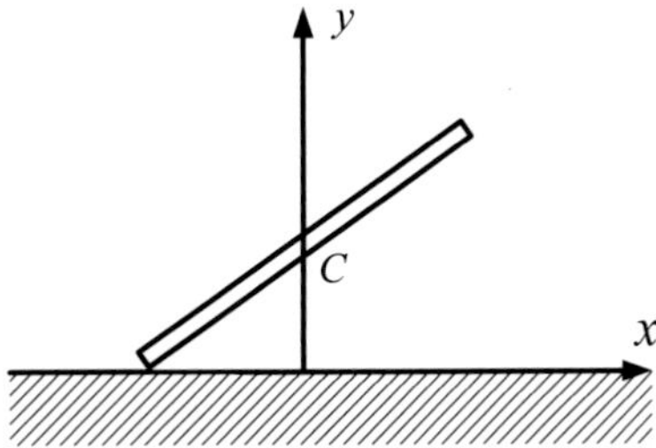

Fig. 11.1 A slender bar sliding on the substrate

However, in the horizontal direction, the resultant force is zero. Therefore, the acceleration of the mass center in this direction is zero. Finally, we can judge that the trajectory of the mass center is a vertical line, for it has displacement in the horizontal direction.

Example 2 As shown in Fig. 11.2, there is a fixed object with the mass center O and gravity $\boldsymbol{P}$. There is a rotator with the gravity $\boldsymbol{p}$, which rotates with respect to point O. The angular displacement is $\varphi = \omega t$, with ω being a constant. Please solve the reaction forces from the fixed substrate.

Answer: The free body diagram is shown in Fig. 11.3. The coordinates of the mass center of the system read

$$\begin{cases} x_C = \dfrac{pe\cos\omega t}{P+p} \\ y_C = \dfrac{pe\sin\omega t}{P+p}. \end{cases}$$

The acceleration of the mass center can be expressed as

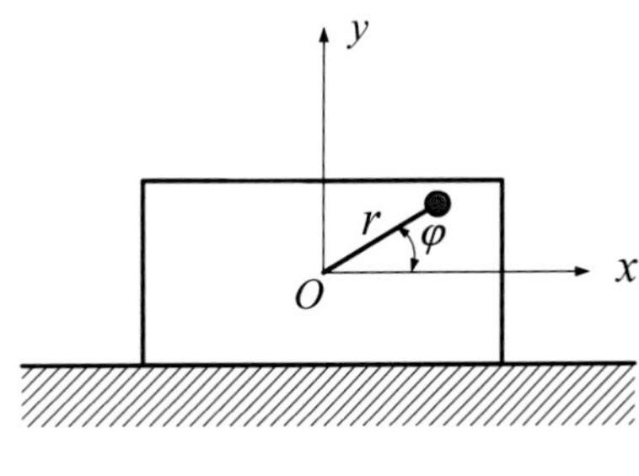

Fig. 11.2 A fixed object and a rotator

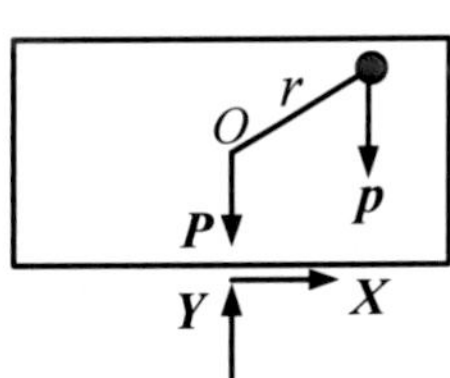

Fig. 11.3 Free body diagram

$$\begin{cases} a_C = -\dfrac{pe\omega^2 \cos \omega t}{P + p} \\ y_C = -\dfrac{pe\omega^2 \sin \omega t}{P + p}. \end{cases}$$

According to the motion theorem of mass center, one has

$$\begin{cases} ma_{Cx} = \Sigma X \\ ma_{Cy} = \Sigma Y. \end{cases}$$

That is

$$\begin{cases} -\dfrac{P + p}{g} \dfrac{pe\omega^2 \cos \omega t}{P + p} = X \\ -\dfrac{P + p}{g} \dfrac{pe\omega^2 \sin \omega t}{P + p} = Y - P - p. \end{cases}$$

Therefore, one has

$$\begin{cases} X = -\dfrac{pe\omega^2}{g} \cos \omega t \\ Y = P + p - \dfrac{pe\omega^2}{g} \sin \omega t. \end{cases}$$

Example 3 As shown in Fig. 11.4, one slender bar rotates with respect to point O, with the length l and mass m. The length of $OA = l/3$. The angular displacement φ, angular velocity ω, and angular acceleration ε are already known. The mass center is located at point C. Please give the reaction forces at point O.

Answer: The free body diagram of the bar is shown in Fig. 11.5.

The acceleration of the mass center is

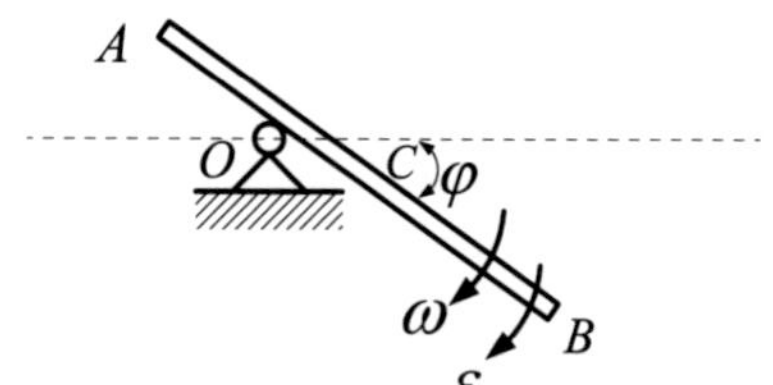

Fig. 11.4 A slender bar rotates to a point

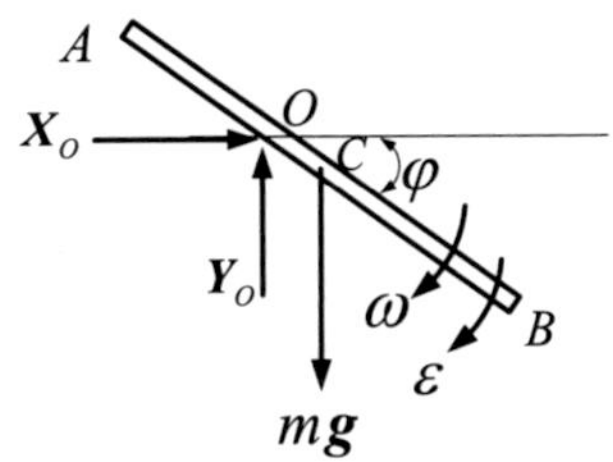

Fig. 11.5 Free body diagram

$$\begin{cases} a_C^n = \dfrac{\omega^2 L}{6} \\ a_C^\tau = \dfrac{\varepsilon L}{6}. \end{cases}$$

The motion theorem of the mass center is

$$\begin{cases} ma_C^n = \Sigma X \\ ma_C^\tau = \Sigma Y. \end{cases}$$

We have established a Cartesian coordinate system along the normal and tangential directions. The equations are

$$\begin{cases} \dfrac{m\omega^2 L}{6} = (Y_O - mg)\sin\varphi - X_O\cos\varphi \\ \dfrac{m\varepsilon L}{6} = (mg - Y_O)\cos\varphi - X_O\sin\varphi. \end{cases}$$

The final solutions are

$$\begin{cases} Y_O = mg + \dfrac{m\omega^2 L}{6}\sin\varphi - \dfrac{m\varepsilon L}{6}\cos\varphi \\ X_O = -\dfrac{m\varepsilon L}{6}\sin\varphi - \dfrac{m\omega^2 L}{6}\cos\varphi. \end{cases}$$

Summary

We should master the knowledge of momentum, theorem of momentum, conservative theorem of the mass center for the rigid body, and motion theorem of the mass center for the rigid body.

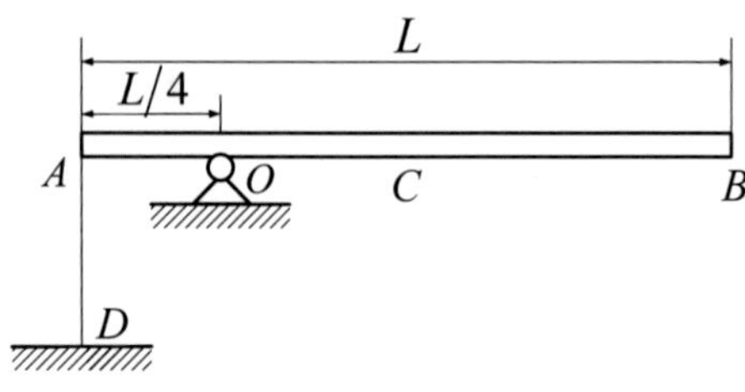

Fig. 11.6 A bar is linked by a rope

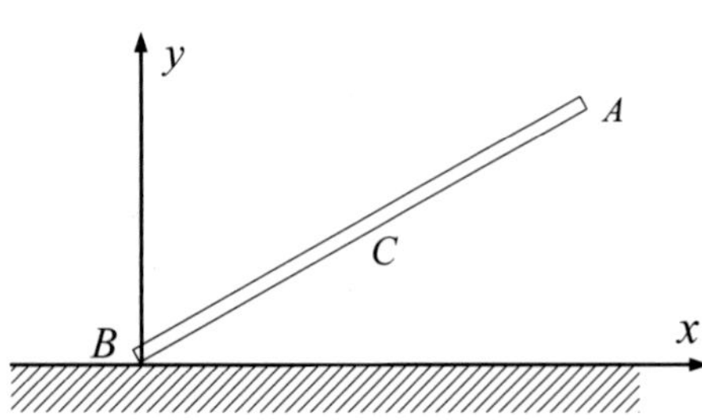

Fig. 11.7 A bar is sliding on a smooth surface

Exercises

11.1 As shown in Fig. 11.6, a slender bar *AB* rotates with respect to point *O*. There is a rope linked at points *A* and *D*. The length of the bar is *L*, mass is *m*, and the centroid is point *C*. When the rope is suddenly cut by a knife, please give the instantaneous angular acceleration of the bar, and the accelerations of point *C* and point *B*.

11.2 As demonstrated in Fig. 11.7, a bar *AB* with the length *L* is contacting with a smooth surface at point *B*, and the angle between the level and the bar is α. Point *C* is the mass center of the bar. Please determine the orbit equation of point *C* in the process of falling with respect to the time *t*.

Answers

11.1 $a_C = \frac{3g}{7}$, $a_B = \frac{9g}{7}$

11.2 $x_c^2 + y_c^2 = \frac{l^2}{4}$

Chapter 12
Theorem of Angular Momentum

Abstract In this chapter, we will first learn the moment of inertia of a rigid body to an axis. Then we will lean the theorem of angular momentum. Especially for a rigid body, the differential equation for the rigid body in rotation is given.

Keywords Moment of inertia · Angular momentum · Theorem of angular momentum · Differential equation for the rigid body in rotation

12.1 Moment of Inertia of a Rigid Body to an Axis

In comparison with the concept of mass, we here first introduce the concept of moment of inertia (or inertial moment), which depicts the rotation property of the rigid body. As shown in Fig. 12.1, a rigid body is rotating with respect to an axis z, with the angular velocity ω. For an arbitrary point m_i, the length between the particle and the axis z is r_i, and velocity is $\boldsymbol{v}_i$. Clearly, each point on the rigid body is in a circular motion, and the velocity is

$$v_i = \omega r_i.$$

The moment of inertia about the rigid body to the axis z is defined as

$$J_z = \sum m_i r_i^2.$$

Clearly, the moment of inertia is always positive, dependent on the shape, axis and mass distribution of the rigid body. It is not related to the motion state of the rigid body. For a rigid body with continuous distribution of mass, it can be further expressed as

$$J_z = \int_V r^2 \mathrm{d}m = \int_V r^2 \rho \, \mathrm{d}V,$$

where V is the volume occupied by the rigid body, and ρ is its density of mass.

J. Liu, *Lecture Notes on Theoretical Mechanics*,
https://doi.org/10.1007/978-981-13-8035-8_12

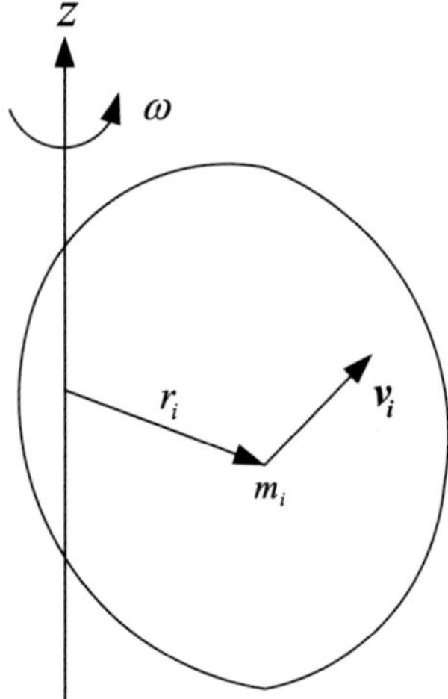

Fig. 12.1 A rigid body rotating to an axis

For some simple rigid bodies, we should remember the formulas of the inertial moment. We assume that z is the axis passing through the mass center. For example, for the disk

$$J_z = \frac{1}{2} m R^2.$$

For the ring, we have

$$J_z = m R^2,$$

where R is the radius of the disk or the ring.

For a slender bar, the moment of inertia is

$$J_z = \frac{1}{12} m L^2.$$

However, for the axis which does not pass through the mass center, how to calculate the moment of inertia? For example, later we shall prove that if the bar rotates with respect to the left end O, then

$$J_O = \frac{1}{3} m L^2.$$

We first introduce the parallel axis theorem. As shown in Fig. 12.2, there are two axes, where z passes through the mass center of the rigid body, and z_1 is parallel to z, with a distance h. The moment of inertia to z-axis is

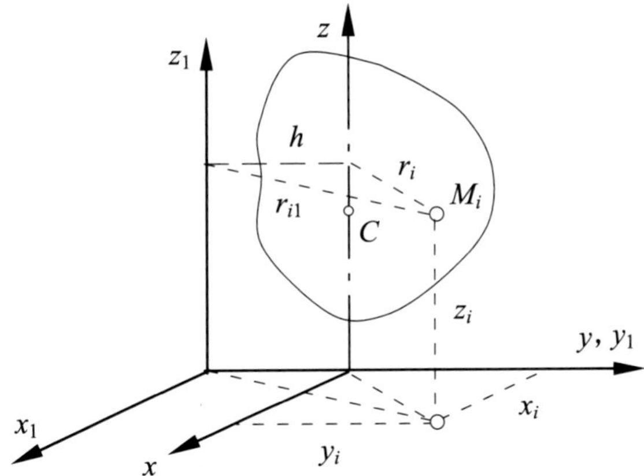

Fig. 12.2 Parallel axis theorem

$$J_z = \sum m_i r_i^2 = \sum m_i (x_i^2 + y_i^2).$$

And the moment of inertia to z_1-axis is

$$J_{z1} = \sum m_i r_{i1}^2 = \sum m_i [x_i^2 + (y_i + h)^2].$$

This equation can be expanded as

$$\begin{aligned} J_{z1} &= \sum m_i [(x_i^2 + y_i^2) - 2hy_i + h^2] \\ &= \sum m_i (x_i^2 + y_i^2) + h^2 \sum m_i - 2h \sum m_i y_i \\ &= J_z + h^2 m - 2hmy_C, \end{aligned}$$

where m is the mass of the rigid body, and y_C is the coordinate of mass center C in the O-xyz system, then $y_C = 0$. Therefore, one has

$$J_{z1} = J_z + h^2 m.$$

For the slender bar, in use of the parallel axis theorem, one has

$$J_O = J_z + m\left(\frac{l}{2}\right)^2 = \frac{ml^2}{12} + \frac{ml^2}{4} = \frac{ml^2}{3}.$$

Example 1 Calculate the moment of inertia on the structure with respect to point O, which is shown in Fig. 12.3. The radius of the disk is R, the length of the bar is L, and O is the midpoint of the bar. The masses of the disk and the bar are, respectively, M and m.

Answer: The moment of inertia with respect to point O includes two portions

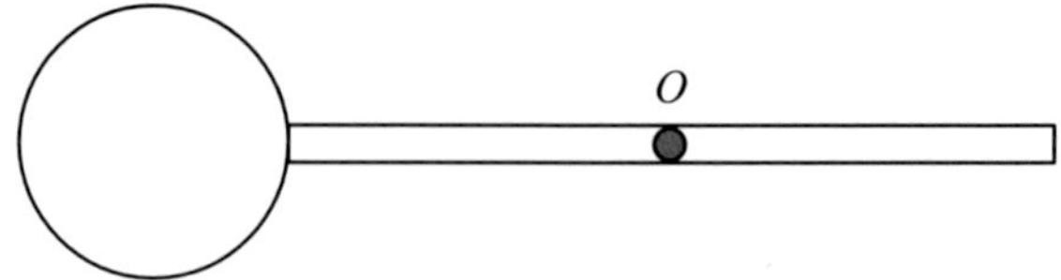

Fig. 12.3 One disk and one bar

$$J_O = J_O^{disc} + J_O^{bar},$$

where

$$J_O^{disc} = \frac{1}{2}MR^2 + M\left(R + \frac{L}{2}\right)^2,$$

$$J_O^{bar} = \frac{1}{12}ML^2.$$

Then the total value is

$$J_O = J_O^{\text{disc}} + J_O^{\text{bar}} = \frac{1}{2}MR^2 + M\left(R + \frac{L}{2}\right)^2 + \frac{1}{12}ML^2.$$

12.2 Theorem of Angular Momentum

The angular momentum of a particle is defined as

$$\boldsymbol{L}_O = \boldsymbol{r} \times \boldsymbol{P} = \boldsymbol{r} \times (m\boldsymbol{v}).$$

Similar to the concept of moment, it can also be decomposed into three components in the three axes, i.e.,

$$\boldsymbol{L}_O = L_x\boldsymbol{i} + L_y\boldsymbol{j} + L_z\boldsymbol{k}.$$

The derivative of the angular momentum is

$$\begin{aligned}\dot{\boldsymbol{L}}_O &= \dot{\boldsymbol{r}} \times m\boldsymbol{v} + \boldsymbol{r} \times m\dot{\boldsymbol{v}} \\ &= \boldsymbol{v} \times m\boldsymbol{v} + \boldsymbol{r} \times m\boldsymbol{a} \\ &= \boldsymbol{r} \times \boldsymbol{F} \\ &= \boldsymbol{M}_O(\boldsymbol{F}).\end{aligned}$$

Similarly, one has

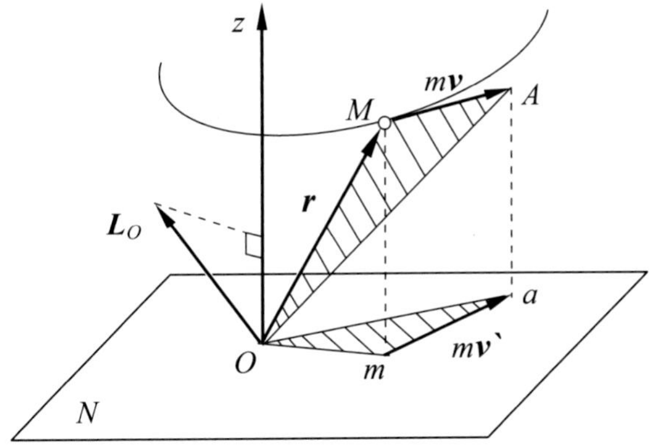

Fig. 12.4 Angular momentum

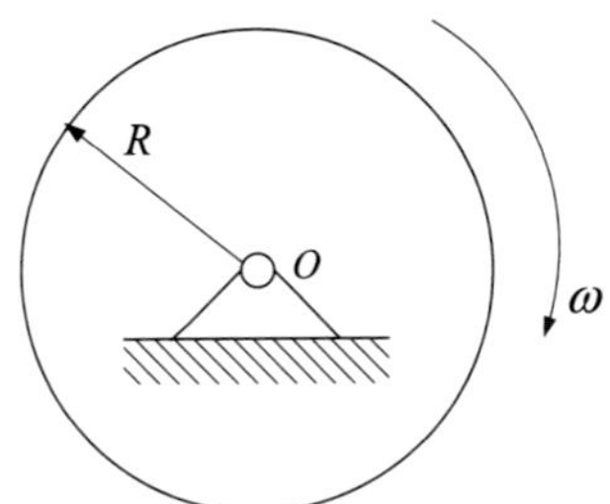

Fig. 12.5 A disk rotating to a point

$$\begin{cases} \dot{L}_x = M_x(\boldsymbol{F}) \\ \dot{L}_y = M_y(\boldsymbol{F}) \\ \dot{L}_z = M_z(\boldsymbol{F}) \end{cases}.$$

As shown in Fig. 12.4, the angular momentum for a rigid body rotating to an axis z is

$$L_z = (\boldsymbol{r} \times m\boldsymbol{v})_z = \sum r_i m_i v_i = \omega \sum m_i r_i^2 = J_z \omega.$$

For example, as shown in Fig. 12.5, a wheel is rotating with respect to a point O. The moment of inertia to point O is

$$J_O = \frac{1}{2} m R^2.$$

Then the angular momentum of the rigid body to point O is

$$L_O = J_O \omega = \frac{1}{2} m R^2 \omega.$$

For a rigid body rotating to an axis z, the theorem of angular momentum is

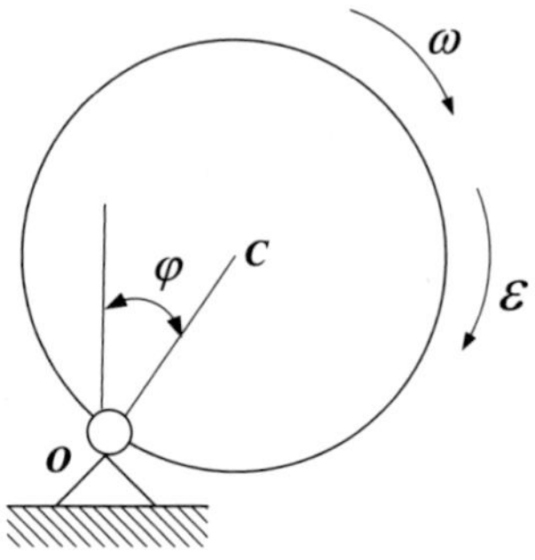

Fig. 12.6 A disk is rotating to a point

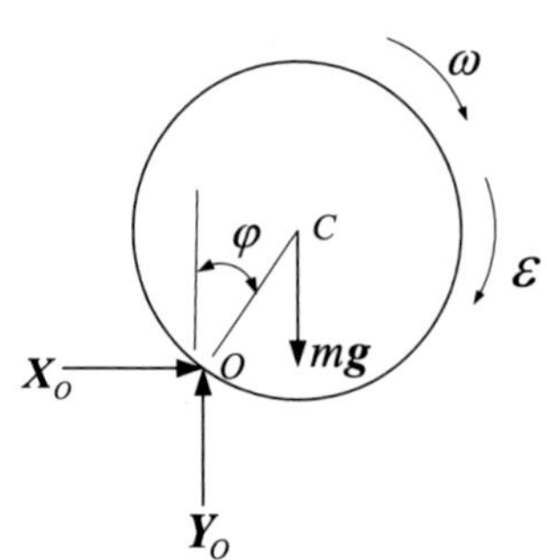

Fig. 12.7 Free body diagram

$$\dot{L}_z = J_z\ddot{\varphi} = J_z\dot{\omega} = J_z\varepsilon = \Sigma M_z(\boldsymbol{F}_i).$$

This means that the variation ratio of the angular momentum is equal to the total moment to this axis.

Example As shown in Fig. 12.6, a disk is rotating with respect to a point O. The mass center of the disk is point C, radius R, mass m, angular velocity ω, and angular acceleration ε. The angle $\varphi = 30°$. Please present the angular momentum to point O and the reactive forces at point O.

Answer: The free body diagram is schematized in Fig. 12.7. The disk is rotating with respect to an axis. According to the theorem, one has

$$J_O\ddot{\varphi} = J_O\dot{\omega} = J_O\varepsilon = \Sigma M_O(\boldsymbol{F}_i),$$

where

$$J_O = \frac{1}{2}mR^2 + mR^2 = \frac{3}{2}mR^2.$$

Therefore, we have

$$\frac{3}{2}mR^2\varepsilon = mgR\,\sin\,\varphi.$$

The angular acceleration is

$$\varepsilon = \frac{2g\sin\varphi}{3R}.$$

The tangential acceleration of the mass center is

$$a_C^{\tau} = \varepsilon R = \frac{2g\sin\varphi}{3}.$$

According to the theorem of motion of the mass center on the rigid body, the equation reads

$$\begin{cases} ma_C^{\tau}\cos\varphi - ma_C^{n}\sin\varphi = X_O \\ ma_C^{\tau}\sin\varphi + ma_C^{n}\cos\varphi = Y_O - mg \end{cases}$$

Therefore, we have

$$\begin{cases} X_O = ma_C^{\tau}\cos\varphi - ma_C^{n}\sin\varphi = \frac{\sqrt{3}}{2}m\varepsilon R - \frac{1}{2}m\omega^2 R \\ Y_O = mg - ma_C^{\tau}\sin\varphi - ma_C^{n}\cos\varphi = mg + \frac{1}{2}m\varepsilon R + \frac{\sqrt{3}}{2}m\omega^2 R \end{cases}$$

The angular momentum is

$$L_O = J_O\omega = \frac{3}{2}mR^2\omega.$$

Summary

In this chapter, we need to first master the moment of inertia of a rigid body to an axis. Then we will master the theorem of angular momentum, especially the differential equation for the rigid body in rotation.

Exercises

12.1 As shown in Fig. 12.8, a curved bar is linked with a disk at point B. The lengths are $AE = BE = BD = R$, where D is the disk center. Both of the bar and the disk have the same mass 3 m. Calculate the moment of inertia respect to point A.

12.2 As shown in Fig. 12.9, a disk is rotating with respect to a point O. The mass center of the disk is C, radius R, mass m, angular velocity ω, and angular acceleration ε. The angle is $\varphi = 30°$. Please present the angular moment to point O and the reactive forces at point O.

12.3 As shown in Fig. 12.10, there is a wheel with radius R. Initially, the wheel is stationary at the step corner, and the angle is $\theta = 0°$. After a minute distur-

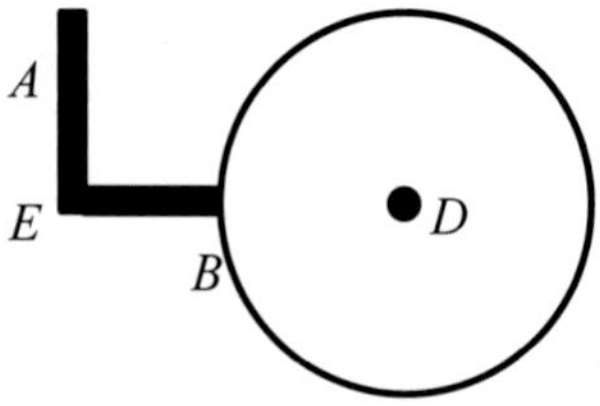

Fig. 12.8 One bar linked with one disk

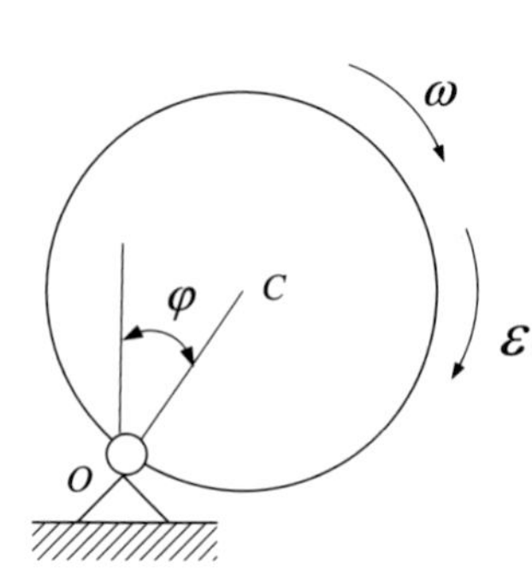

Fig. 12.9 A disk is rotating with respect to point O

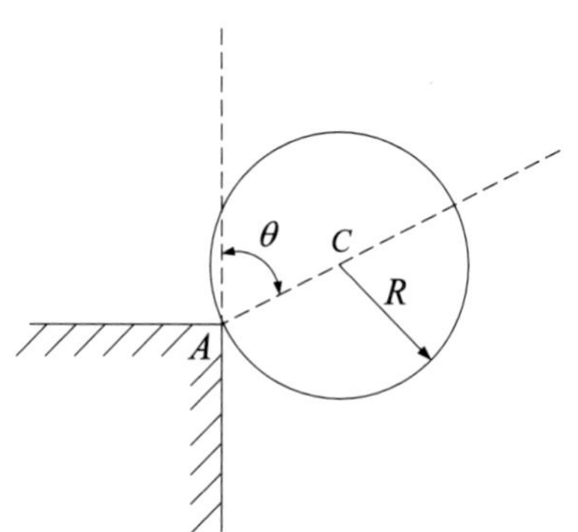

Fig. 12.10 A roller at the step corner

bance, the wheel will fall down on the contact surface in pure rolling. Please calculate the critical angle θ and critical rotation speed n when the wheel just leaves the step.

12.4 As shown in Fig. 12.11, a ring C is fixed with a bar, and they are of the same mass m. The ring is linked by a rope AB, and the angle between the bar and the horizontal direction is $\varphi = 30°$. The radius of the ring is R, and the length of the bar is $3R$. Suddenly, when the rope is sheared, please calculate the following variables:

(1) The moment of inertia of the structure with respect to point O.
(2) The angular acceleration of this structure.
(3) The reactive forces at point O.

12.5 As shown in Fig. 12.12, the oblique surface is fixed. There is a slider C with gravity Q sliding upward along the surface. The slider is linking through a

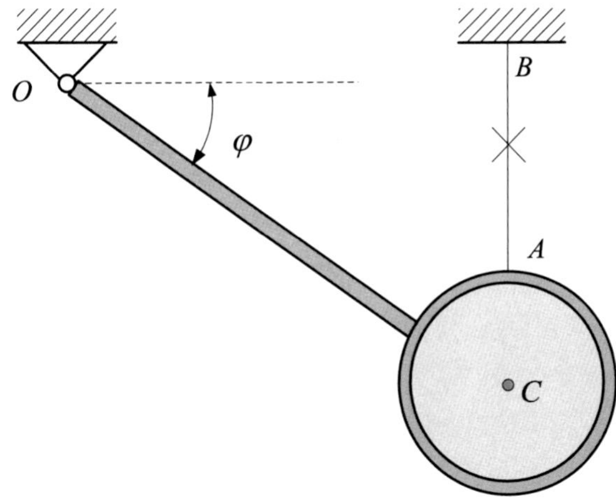

Fig. 12.11 A structure linked by a rope suddenly sheared

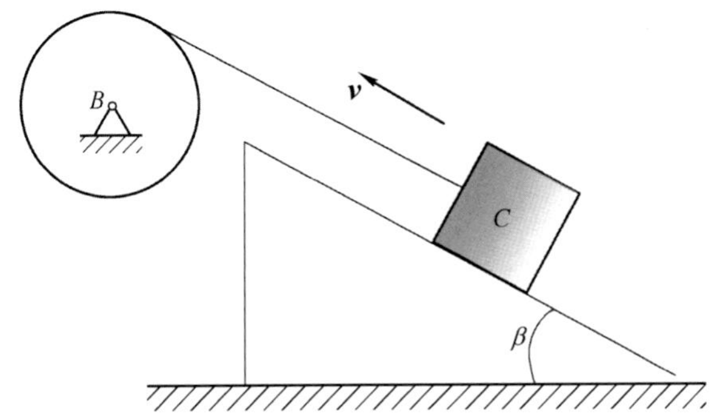

Fig. 12.12 A slider driven by a wheel

rope on the edge of a fixed wheel. The radius of the wheel is R, and gravity is W. At the current time, the velocity of the slider is $\boldsymbol{v}$. Please write down the expressions of momentum, angular momentum to point B, and kinetic energy.

Answers

12.1 $19\,mR^2$

12.2 $J_O = \frac{3}{2}mR^2$, $F_{Ox} = \frac{\sqrt{3}}{2}mR\varepsilon - \frac{1}{2}mR\omega^2$, $F_{Oy} = -\frac{1}{2}mR\varepsilon - \frac{\sqrt{3}}{2}mR\omega^2$

12.3 $\theta = \arccos\frac{4}{7}$, $n = \frac{1}{\pi}\sqrt{\frac{g}{7r}}$

12.4 (1) $J_{\text{o}} = 20mR^2$ (2)$\varepsilon = \frac{11g\cos\varphi}{40R}$
Along the direction of the bar $N_1 = 2mg\sin\varphi$.
Vertical direction of the bar $N_2 = \frac{281}{80}mg\cos\varphi$.

12.5 $P = \frac{Q}{g}v$, $L_{\mathbf{B}} = \frac{mvR^2}{3}$

Chapter 13
Theorem of Kinetic Energy

Abstract In this chapter, we will learn the theorem of kinetic energy. First, the concepts of work and kinetic energy are given. Then, the theorem of kinetic energy is introduced, and some examples are given to show the solving steps.

Keywords Work · Kinetic energy · Theorem of kinetic energy

The concept of energy was first proposed by T. Young, a great physician, mechanics, and physics scholar, who introduced the concept of Young's modulus and theory of two-slit interference of light. The work can be transferred into energy, so these two types of physical quantities must have close relations. We, here, first introduce the concept of work.

13.1 Work

"Work" is defined as the cross product of the force and the corresponding displacement, which is formulated as

$$\begin{aligned} W &= \boldsymbol{F} \cdot \boldsymbol{r} \\ &= Fs \cos \varphi, \end{aligned}$$

where $\boldsymbol{F}$ is a constant force, s is the magnitude of the displacement, and φ is the angle between the lines of the applied force and the displacement, as shown in Fig. 13.1. This means that the work represents the actual contribution of the force to the displacement, and it can be understood that the displacement times the projection quantity of the force along the line of displacement. Notably, the work is a scalar, whose unit is J, which is in memory of the great scientist Joule.

For an arbitrary force, the work should be expressed via integration

$$W = \int_C \boldsymbol{F} \cdot \mathrm{d}\boldsymbol{r}$$

J. Liu, *Lecture Notes on Theoretical Mechanics*,
https://doi.org/10.1007/978-981-13-8035-8_13

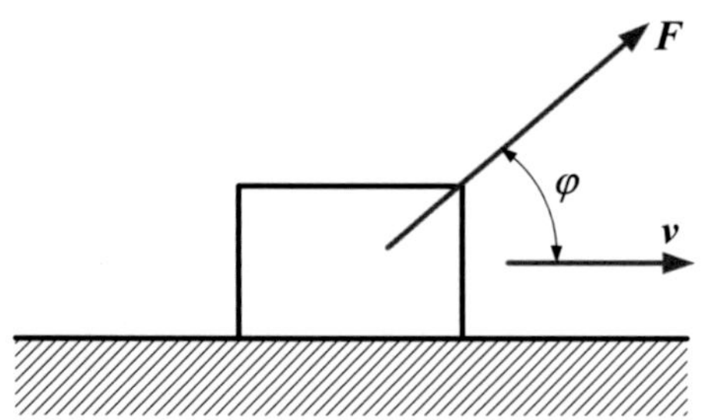

Fig. 13.1 Force and displacement

$$= \int_C F \cos \varphi \mathrm{d}s$$
$$= \int_C Fv \cos \varphi \mathrm{d}t.$$

Let the displacement and force be

$$\boldsymbol{r} = x\boldsymbol{i} + y\boldsymbol{j} + z\boldsymbol{k},$$

$$\boldsymbol{F} = X\boldsymbol{i} + Y\boldsymbol{j} + Z\boldsymbol{k},$$

respectively, then the work can be rewritten as

$$W = \int_C (X\mathrm{d}x + Y\mathrm{d}y + Z\mathrm{d}z)$$
$$= \int_C \left(Xv_x + Yv_y + Zv_z\right)\mathrm{d}t$$

For the rigid body subjected to the work from gravity, the work is

$$W = mgh_C,$$

where h_C is the height difference of the mass center of the rigid body, m is the mass and g is the gravitational acceleration. This work is not related with the initial and final height but with the height difference of the mass center only.

Let's then look at the work done by the spring, in connection with a mass, which is shown in Fig. 13.2. The elongation of the mass is x from the equilibrium point, then the force produced by the spring is

$$F = -kx,$$

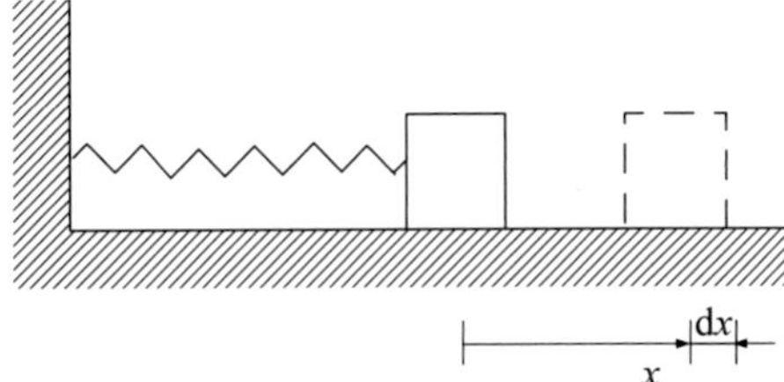

Fig. 13.2 Spring

where k is the stiffness coefficient of the spring. The work done from two points, with elongations from λ_1 to λ_2 is

$$W = \int_{\lambda_1}^{\lambda_2} F\mathrm{d}x$$

$$= -\int_{\lambda_1}^{\lambda_2} kx\mathrm{d}x$$

$$= \frac{1}{2}k\left(\lambda_1^2 - \lambda_2^2\right).$$

For a rigid body in translation, the work is

$$W = \int_C \boldsymbol{F} \cdot \mathrm{d}\boldsymbol{r}_C.$$

However, for a rigid body in rotation, the work done is caused by a moment, and the effect is the rotation of the angle, so the work is

$$W = \int_0^{\varphi} M_z \mathrm{d}\varphi.$$

Throughout this textbook, we only consider the ideal constraint, which has no work. That means, this reactive force from the ideal constraint has no contribution to the displacement.

13.2 Kinetic Energy

The concept of kinetic energy was initially not understood clearly, which was considered not distinct with the concept of momentum. The scientist Descartes proposed that the energy should be defined as $m\boldsymbol{v}$, and we now know that it is a vector and not proper. Later, another scientist Leibniz gave the correct definition, which is a scalar. For a particle, the kinetic energy can be defined as

$$T = \frac{1}{2}mv^2.$$

For a particle system which embraces a lot of particles, its kinetic energy is the summation of all particles, which is expressed as

$$T = \sum \frac{1}{2}m_i v_i^2,$$

where i denotes the order of the particle.

However, in dynamics, we are mainly concerned with the rigid body. For a rigid body in translation, all the particles have the same velocity. We use the information of one characteristic point to denote the velocity, and choose the mass center to examine. The kinetic energy is

$$\begin{aligned} T &= \sum \frac{1}{2}m_i v_i^2 = \frac{1}{2}\left(\sum m_i\right) v_C^2 \\ &= \frac{1}{2}mv_C^2, \end{aligned}$$

where m is the total mass of the rigid body.

For a rigid body in rotation, an arbitrary point on the rigid body is in a circular motion. As shown in Fig. 13.3, its velocity is

$$v_i = \omega r_i.$$

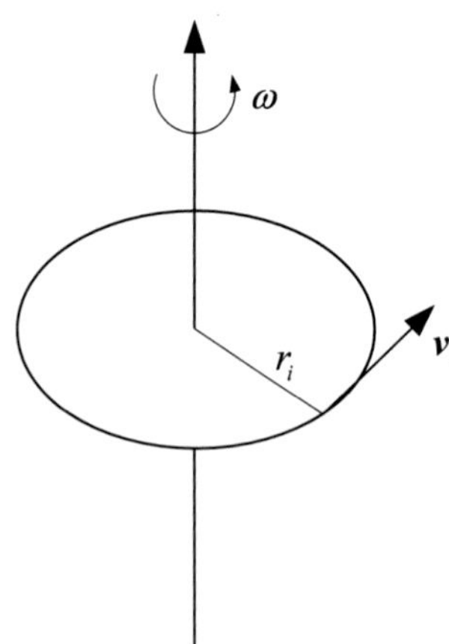

Fig. 13.3 Rotation

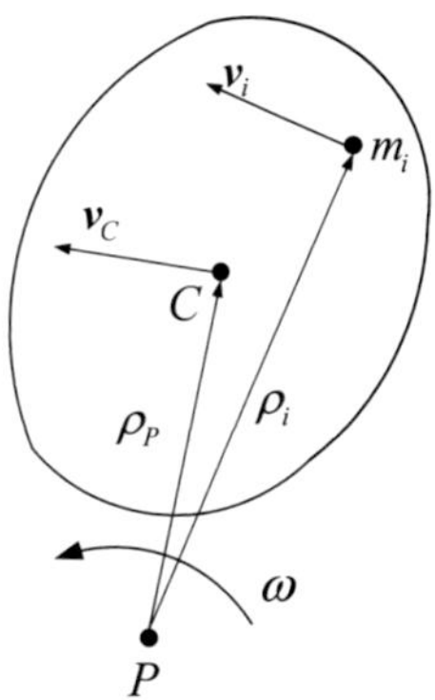

Fig. 13.4 A rigid body in planar motion

Then, the kinetic energy can be expressed as

$$T = \sum \frac{1}{2} m_i v_i^2 = \frac{1}{2}\left(\sum m_i r_i^2\right)\omega^2$$
$$= \frac{1}{2} J_z \omega^2,$$

where $J_z = \sum m_i r_i^2$ is the moment of inertia on the z-axis.

The most complex case is the planar motion of a rigid body, whose instantaneous center is P, and the angular velocity is ω, as shown in Fig. 13.4. The center of mass is point C, with the distance $CP = \rho_P$. An arbitrary point with the mass m_i takes the distance from the instantaneous center of ρ_i. The velocity of this point is

$$v_i = \omega \rho_i,$$

and the velocity of the mass center is

$$v_C = \omega \rho_P.$$

The kinetic energy can be expressed as

$$T = \sum \frac{1}{2} m_i v_i^2 = \frac{1}{2}\left(\sum m_i \rho_i^2\right)\omega^2$$
$$= \frac{1}{2} J_P \omega^2,$$

where $J_P = \sum m_i \rho_i^2$ is the moment of inertia on point P. However, the instantaneous center will vary at different time, so it is not a constant, and this will bring the more complicated calculations. We then think of a new expression of the kinetic energy, in seek of another characteristic point, i.e., the center of mass C. In use of the translation law of the parallel axis, one has

$$
\begin{aligned}
J_P &= J_C + mCP^2 \\
&= J_C + m\rho_P^2.
\end{aligned}
$$

Consequently, one has

$$
\begin{aligned}
T =& \frac{1}{2} J_P \omega^2 \\
=& \frac{1}{2} J_C \omega^2 + \frac{1}{2} m \left(\omega \rho_P^2\right)^2 \\
=& \frac{1}{2} m v_C^2 + \frac{1}{2} J_C \omega^2.
\end{aligned}
$$

This reemphasizes that the planar motion of the rigid body can be viewed that it includes both translation and rotation. Therefore, the expression of the kinetic energy in planar motion is naturally decomposed of two terms, in which one represents translation and the other represents rotation.

13.3 Theorem of Kinetic Energy

The theorem of kinetic energy aims at building the relation between the work and the kinetic energy. We first derive this theorem from a particle. The basic origin is the Newton's second law, which reads

$$
m\boldsymbol{a} = m\dot{\boldsymbol{v}} = \boldsymbol{F}.
$$

Multiplying the velocity $\boldsymbol{v}$ to both sides of the above equation, one has

$$
m\dot{\boldsymbol{v}} \cdot \boldsymbol{v} = \boldsymbol{F} \cdot \boldsymbol{v}.
$$

It can be recast into

$$
m\boldsymbol{v} \cdot \mathrm{d}\boldsymbol{v} = \boldsymbol{F} \cdot \boldsymbol{v}\mathrm{d}t.
$$

Integration of the above equation yields

$$
T_2 - T_1 = W,
$$

where $T = \frac{1}{2}mv^2$, $W = \int_C \boldsymbol{F} \cdot \mathrm{d}\boldsymbol{r}$, and T_1 and T_2 are the kinetic energies of the initial and final states of the particle. For a rigid body, the expression of this theorem can be given according to the motion types. It should be mentioned that, if we use the theorem of kinetic energy of a particle system, we should write down that $T = \sum \frac{1}{2} m_i v_i^2$.

The above equation is the integration format of the theorem of kinetic energy, and we have another format by differentiating it

$$\mathrm{d}T = \mathrm{d}W.$$

As a consequence, one can see that based on the theorem of kinetic energy, we can solve a lot of physical variables. For instance, the force, displacement, mass, velocity, moment, and angular displacement, angular velocity, and moment of inertia can be all solved by the integration format. From the differential format, one can solve the acceleration, velocity, angular velocity, and angular acceleration.

Here are the steps about how to use the theorem of kinetic energy:

(1) Select the object, and isolate it.
(2) Perform force analyses. Analyze the active forces and reactive forces, and then draw the free body diagram.
(3) Analyze the work caused by the forces. Judge which forces do work, and which forces do not work. Then, judge the sign of the work. Notice the work done by the constraint.
(4) Write out the initial and final expressions of the kinetic energy.
(5) Apply the integration format of the theorem.
(6) Apply the differential format of the theorem.
(7) Get the solution.

Example 1: A slender bar is lying on the wall corner, and the two surfaces are both smooth. The length of the bar is L, the velocity $\boldsymbol{v}_A = \boldsymbol{u}$. Please write out the expression of the kinetic energy.

Answer: We first find the instantaneous center of the bar, which is shown in Fig. 13.5. We have the velocities as

$$v_A = \omega AP = u,$$

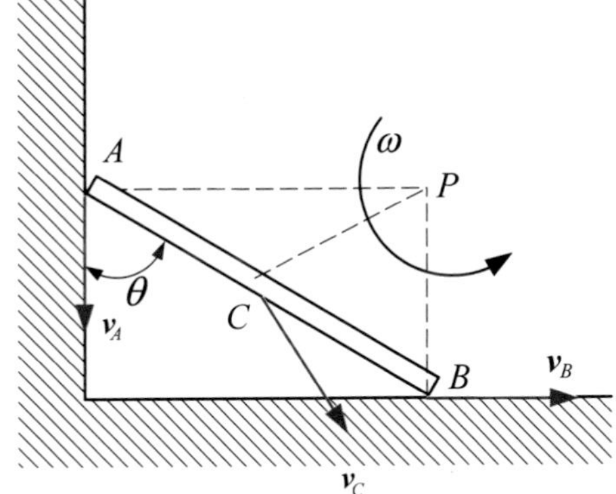

Fig. 13.5 A bar at the corner

$$v_B = \omega BP,$$

$$v_C = \omega CP.$$

The geometric relations are

$$AP = L \sin \theta,$$

$$BP = L \cos \theta,$$

$$CP = \frac{L}{2}.$$

Then, the angular velocity is

$$\omega = \frac{u}{L \sin \theta},$$
$$v_C = \frac{u}{2 \sin \theta}.$$

The moment of inertia is

$$J_C = \frac{mL^2}{12}.$$

The bar is in planar motion, whose kinetic energy is

$$\begin{aligned} T =& \frac{1}{2} m v_C^2 + \frac{1}{2} J_C \omega^2 \\ =& \frac{mu^2}{6 \sin^2 \theta}. \end{aligned}$$

Example 2: A disk is purely rolling on an inclined surface with the angle of α, as shown in Fig. 13.6. The gravity of the disk is mg, and the mass center is point C. Initially, the disk was static, and some time later, it will move along the surface, with the displacement s. The radius of the disk is R. Please solve the velocity and acceleration of the mass center, the angular velocity, and angular acceleration of the disk.
Answer: The disk is in planar motion. We select the disk as the object, and perform the force analysis: the active force $\boldsymbol{mg}$, the normal contact force $\boldsymbol{N}$, and the friction force $\boldsymbol{F}$. The constraint is an ideal constraint, so there is no work from the contact force and the friction force. Therefore, the work in the process is

$$W = mgs \sin \alpha.$$

The initial kinetic energy is

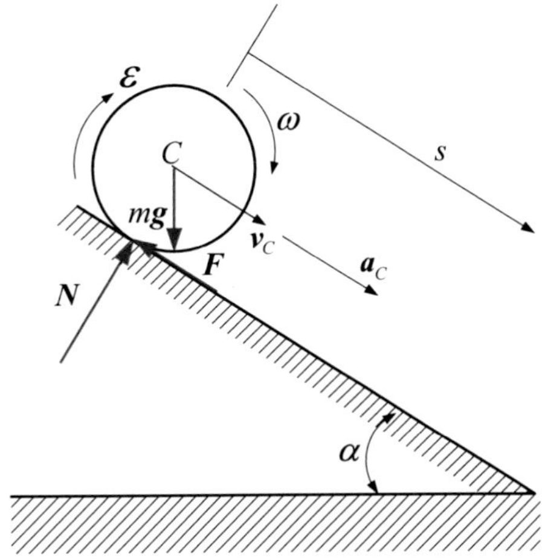

Fig. 13.6 A disk on an inclined surface
（图中应为小写）

$$T_1 = 0.$$

The final kinetic energy is

$$T_2 = \frac{1}{2}mv_C^2 + \frac{1}{2}J_C\omega^2.$$

The disk is in pure rolling, so the contact point with the inclined surface is the instantaneous center. We then have

$$v_C = \omega R,$$

$$J_C = \frac{1}{2}mR^2.$$

The final kinetic energy can be recast into

$$T_2 = \frac{3}{4}mR^2\omega^2.$$

The theorem of kinetic energy is

$$T_2 - T_1 = W.$$

Then one has

$$\frac{3}{4}mR^2\omega^2 - 0 = mgs\,\sin\alpha.$$

The angular velocity is

$$\omega = \frac{2}{R}\sqrt{\frac{g\ \sin\ \alpha}{3}}.$$

The velocity of the mass center is

$$v_C = \omega R = 2\sqrt{\frac{g\ \sin\ \alpha}{3}}$$

Taking derivative to both sides of the equation about the theorem of kinetic energy, one has

$$a_C = \varepsilon R.$$

The differential format of the theorem of kinetic energy is

$$\mathrm{d}T = \mathrm{d}W,$$

$$\frac{3}{2}R^2\omega\dot{\omega} = g\ \sin\ \alpha\dot{s},$$

$$\frac{3}{2}R^2\omega\varepsilon = g\ \sin\ \alpha v_C = g\ \sin\ \alpha\omega R$$

Then, the angular acceleration is

$$\varepsilon = \frac{2g\ \sin\ \alpha}{3R}.$$

The acceleration of the mass center is

$$a_C = \varepsilon R = \frac{2g\ \sin\ \alpha}{3}.$$

Example 3: A wheel is fixed with a cylindrical hinge on the ground, driven by a motor providing a constant moment M. There is a rope coiling on the wheel, with a rigid body, as shown in Fig. 13.7. The wheel has rotated with an angle of φ, and the mass of the object hanging on the rope is m. The radius of the wheel is R. Please present the angular velocity and angular acceleration of the wheel when the angle φ is given.
Answer: The work done by the moment and gravity of the object is

$$W = M\varphi - mgh,$$

where h is the height elevation of the object, which satisfies

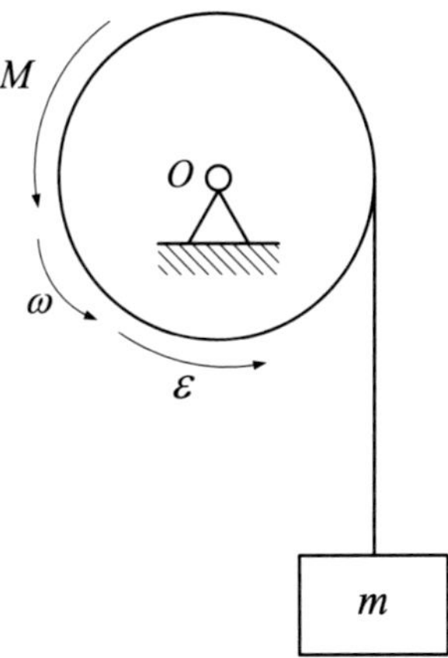

Fig. 13.7 A wheel–mass system

$$h = \varphi R.$$

Then, the work can be written as

$$W = (M - mgR)\varphi.$$

The initial kinetic energy of the system is

$$T_1 = 0.$$

The final kinetic energy of the system is

$$T_2 = \frac{1}{2}J_O\omega^2 + \frac{1}{2}mv^2,$$

where v is the velocity magnitude of the object, ω the angular velocity of the wheel, and the moment of inertia on point O is

$$J_O = \frac{1}{2}mR^2.$$

The velocity of the object is equal to the velocity of a point on the wheel edge, since the rope is assumed inextensible, so one has

$$v = \omega R.$$

The final kinetic energy of the system can be recast into

$$T_2 = \frac{1}{2}J_O\omega^2 + \frac{1}{2}mv^2$$

$$=\frac{3}{4}mR^2\omega^2.$$

The theorem of kinetic energy is

$$T_2 - T_1 = W.$$

Then one has

$$\frac{3}{4}mR^2\omega^2 - 0 = (M - mgR)\varphi.$$

The angular velocity is

$$\omega = \frac{2}{R}\sqrt{\frac{(M - mgR)\varphi}{3m}}.$$

The velocity of the object is

$$v = \omega R = 2\sqrt{\frac{(M - mgR)\varphi}{3m}}.$$

The differential format of the theorem of kinetic energy is

$$\mathrm{d}T = \mathrm{d}W,$$

$$\frac{3}{2}mR^2\omega\dot{\omega} = (M - mgR)\dot{\varphi},$$

$$\frac{3}{2}mR^2\omega\varepsilon = (M - mgR)\omega.$$

The angular acceleration of the wheel can be given as

$$\varepsilon = \frac{2(M - mgR)}{3mR^2}.$$

The acceleration of the object is

$$a = \varepsilon R = \frac{2(M - mgR)}{3mR}.$$

Summary

In this chapter, we must master the theorem of kinetic energy, which is a powerful tool to solve problems in dynamics. First, we should master the concepts of work and kinetic energy. Then, we should master the steps on how to use the theorem of kinetic energy to solve problems in dynamics.

Exercises

13.1 As shown in Fig. 13.8, a roller is purely rolling from the oblique surface with an angle α. The mass of the disk is m, radius R, and the displacement of the mass center is s. At this time, please solve the velocity and acceleration of the mass center C, angular velocity, and angular acceleration of the disk.

13.2 As shown in Fig. 13.9, a roller is purely rolling on the horizontal surface, with the radius R. The smaller wheel has no mass, with the radius r, and it rotates with respect to point B. The gravities of the weight A and the roller are both $\boldsymbol{Q}$. The center of the roller and the weight are linked by an inextensible rope, and the rope spans the smaller wheel. Initially, the system was stationary. When the weight A moves vertically with a displacement s, please solve the velocity and acceleration of the weight A.

13.3 As shown in Fig. 13.10, a panel C with mass M is linking the mass center of the wheel A, and contacting with a roller B. The panel is always parallel to the ground in the motion process. The mass of the wheel A is m, and its radius is r. The mass of the roller B is also m, and its radius is R. The wheel A has no sliding with the ground, and the panel has also no sliding with the roller B. The velocity of the panel is v. Please give the expressions of the momentum and kinetic energy of the system.

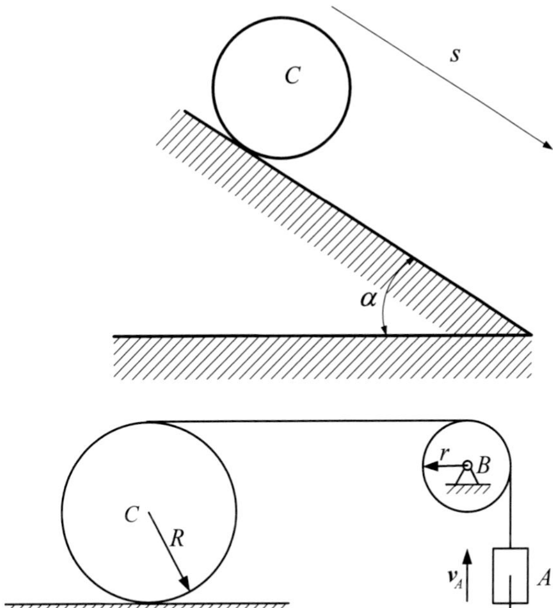

Fig. 13.8 A roller is rolling on an oblique surface

Fig. 13.9 Two rollers and one weight

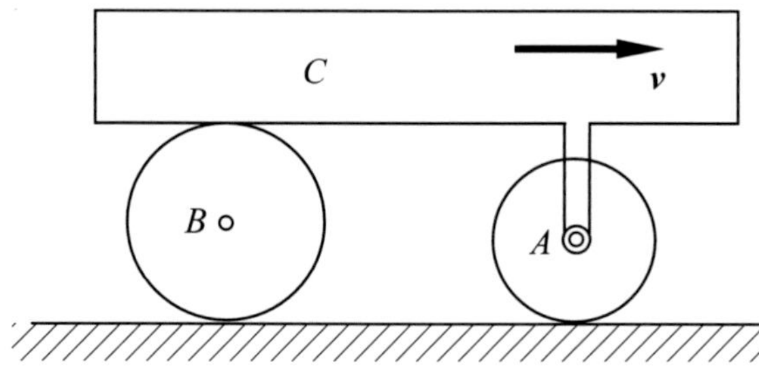

Fig. 13.10 A panel contacting a wheel and a roller

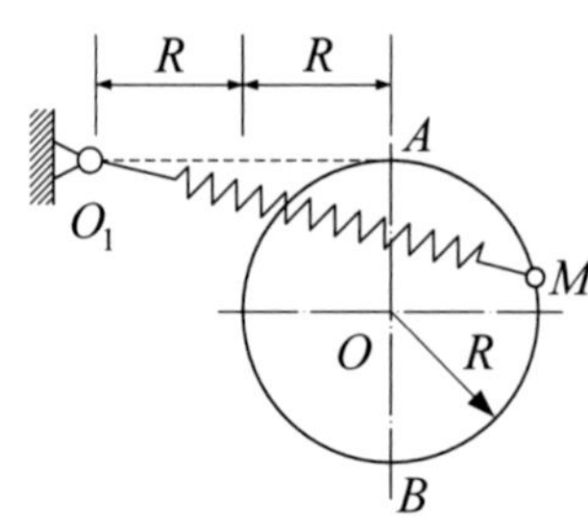

Fig. 13.11 A spring linked by a ring

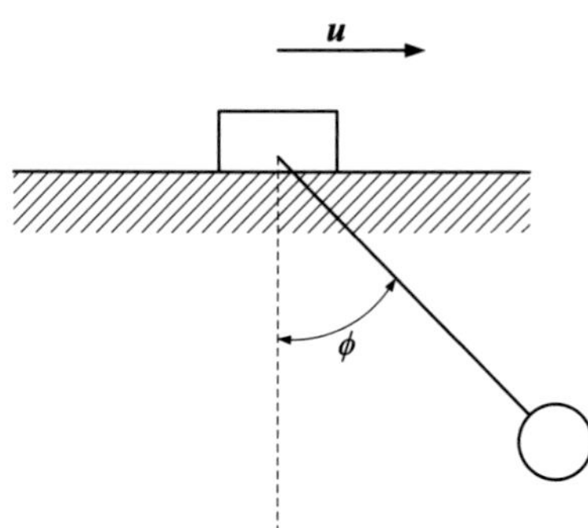

Fig. 13.12 A pendulum linking a slider

13.4 As shown in Fig. 13.11, there is a spring with the elastic constant *k*, and its natural length is *R*. The left end is fixed at point O_1, and another end is linked at a smaller ring *M*. The center of the bigger ring is fixed at point *O*. The smaller ring *M* is in a circular motion on the fixed ring with radius *R*. In the process when the ring *M* moves from point *A* to point *B*, please calculate the work done by the spring.

13.5 As shown in Fig. 13.12, a slider with mass *M* is moving on the horizontal surface with a uniform velocity $\boldsymbol{u}$. There is a pendulum linking the slider at point *O*, where its length is *L*, and mass is *m*. The rotation equation of the pendulum is $\phi = \phi(t)$. Then, write down the expressions of the momentum and kinetic energy of the system.

Answers

13.1 $v_c = \frac{2\sqrt{3sg\ \sin\alpha}}{3}$, $a_c = \frac{2g\ \sin\alpha}{3}$
13.2 $v_c = \frac{2\sqrt{5sg}}{5}$, $a_c = \frac{2g}{5}$
13.3 $\left(M + \frac{3}{2}m\right)v$
13.4 $2\left(3 - 2\sqrt{2}\right)kR^2$
13.5 $\frac{1}{2}(M + m)u^2 + \frac{1}{2}mL^2\dot{\phi}^2 + muL\dot{\phi}\cos\phi$

Chapter 14
Lagrange Equation

Abstract In this chapter, we will learn the knowledge on Lagrange equation. We mainly learn how to use it to solve problems in dynamics, and some examples are given.

Keywords Lagrange equation · Conservative force · Degree of freedom

In engineering, scientists and engineers are often faced with very complicated systems, which can be viewed as a particle system. We consider one particle system including n particles, and the active force acting on each particle is $\boldsymbol{F}_i$, and its displacement vector is denoted by $\boldsymbol{r}_i$. Evidently, to solve this kind of problem is not trivial. For each particle, we can get one motion equation according to the Newton's second law, where the reactive forces among particles must be considered. For example, if there are hundreds of particles inside one system, we must solve the equation group containing hundreds of coupled equations. For the calculation by hand, it is almost impossible. How to solve this kind of complex problem fosters people to find new methods and new theories, and Lagrange made great contributions in this area. After Euler, Lagrange became the greatest mathematician all over the world. One of his previous students, Napoléon Bonaparte praised Lagrange as "A high pyramid in mathematics".

In his celebrated book "Analytical Mechanics", the great scientist Lagrange derived the famous Lagrange equation

$$\frac{\mathrm{d}}{\mathrm{d}t}\left(\frac{\partial T}{\partial \dot{q}_j}\right) - \frac{\partial T}{\partial q_j} = Q_j \quad (j = 1, 2, \ldots, k).$$

The full name of this equation is the Lagrange equation of second class. In this equation, Q_j is the generalized force, which is defined as

$$Q_j = \sum_{i=1}^{n} \boldsymbol{F}_i \cdot \frac{\partial \boldsymbol{r}_i}{\partial q_j},$$

where q_j is the generalized coordinate, which is used to describe the position for each particle. The kinetic energy for the system is defined as

J. Liu, *Lecture Notes on Theoretical Mechanics*,
https://doi.org/10.1007/978-981-13-8035-8_14

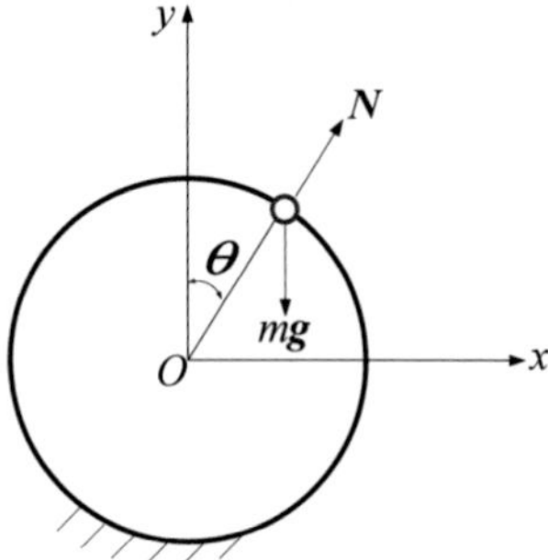

Fig. 14.1 A small ring sliding on a smooth fixed ring

$$T = \sum_{i=1}^{n} \frac{1}{2} m_i v_i^2.$$

In particular, for a conservative system, the generalized force can be expressed as the derivative of the potential energy, i.e.,

$$Q_j = -\frac{\partial U}{\partial q_j}.$$

It should be mentioned that the potential energy is not correlated with the generalized velocity, and thus, the Lagrange equation for the conservative system can be recast as

$$\frac{\mathrm{d}}{\mathrm{d}t}\left(\frac{\partial T}{\partial \dot{q}_j}\right) - \frac{\partial T}{\partial q_j} + \frac{\partial U}{\partial q_j} = 0.$$

Introducing the Lagrange function

$$L = T - U.$$

One can get the expression of this equation

$$\frac{\mathrm{d}}{\mathrm{d}t}\left(\frac{\partial L}{\partial \dot{q}_j}\right) - \frac{\partial L}{\partial q_j} = 0 \quad (j = 1, 2, \ldots, k).$$

We can find that there are no reactive forces in this equation, and it is convenient to solve complex dynamic problems.

Example 1: As shown in Fig. 14.1, a small ring is sliding on a bigger ring fixed on the ground. The mass of the small ring is m, and the radius of the bigger ring is r. Please build the differential equations of the small ring.

Answer: The Cartesian coordinate system O-xy is shown in Fig. 14.1. We select the angle θ as the generalized coordinate. The kinetic energy is written as

$$T = \frac{1}{2} m r^2 \dot{\theta}^2,$$

and the potential energy due to gravity is

$$U = mgr \cos\theta.$$

Therefore, the Lagrange equation is

$$\begin{aligned} L &= T - U \\ &= \frac{1}{2} mr\left(r\dot{\theta}^2 - 2g\cos\theta\right). \end{aligned}$$

The derivatives are

$$\begin{aligned} \frac{\partial L}{\partial \dot{\theta}} &= mr^2\dot{\theta}, \\ \frac{\partial L}{\partial \theta} &= mgr\sin\theta. \end{aligned}$$

Substituting the above two equations into the Lagrange equation, one has

$$mr^2\ddot{\theta} - mgr\sin\theta = 0,$$

i.e.,

$$\ddot{\theta} - \frac{g}{r}\sin\theta = 0.$$

In fact, this equation can also be obtained according to the Newton's second law. As shown in Fig. 14.1, there is one gravitational force and one reactive force $\boldsymbol{N}$ acting on the ring. Consequently, in the tangential direction, one can get

$$m\frac{\mathrm{d}v}{\mathrm{d}t} = mgr\sin\theta,$$

where

$$\frac{\mathrm{d}v}{\mathrm{d}t} = \frac{\mathrm{d}\left(r\dot{\theta}\right)}{\mathrm{d}t} = r\ddot{\theta}.$$

Finally, we can get the same result as that from the Lagrange equation. This again verifies the consistence between the Newton's equation and Lagrange equation, from different viewpoints.

Example 2: As shown in Fig. 14.2, a planet gear system incorporates a smaller gear and a bigger gear. The mass of the smaller gear is m_2, and radius is r, which is viewed

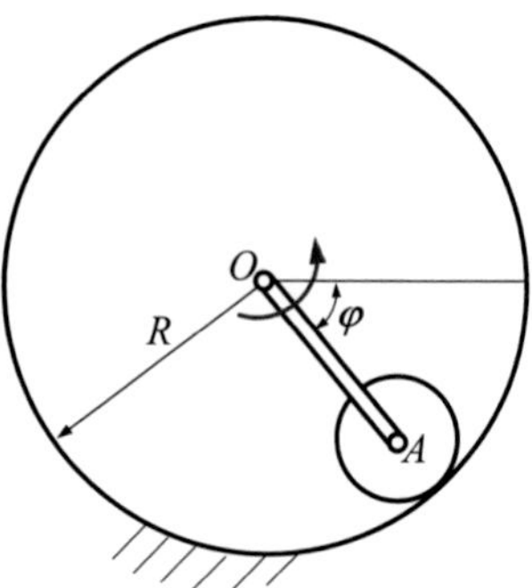

Fig. 14.2 A planet gear system

as a circular disk. The bar linking the small gear has the mass m_1, and the radius of the bigger gear is R. There is a couple acting M on the bar, and the smaller gear sliding on the bigger gear. Please give the expression of the angular velocity of the bar OA.

Answer: We select the angle φ as the generalized coordinate. Obviously, this is not a conservative system. As a result, the work from the nonconservative force is

$$W = -M\varphi.$$

The potential energy is

$$U = 0,$$

And, the kinetic energy is expressed as

$$\begin{aligned} T &= \frac{1}{2}\frac{m_1}{3}(R-r)^2\dot{\varphi}^2 + \frac{1}{2}\left(\frac{m_2}{2}r^2 + m_2r^2\right)\left(\frac{R-r}{r}\right)^2\dot{\varphi}^2, \\ &= \frac{1}{12}(2m_1 + 9m_2)(R-r)^2\dot{\varphi}^2. \end{aligned}$$

The Lagrange equation is

$$\begin{aligned} L &= T - U, \\ &= \frac{1}{12}(2m_1 + 9m_2)(R-r)^2\dot{\varphi}^2. \end{aligned}$$

The generalized force is derived as

$$Q = \frac{\mathrm{d}W}{\mathrm{d}\varphi} = -M.$$

According to the Lagrange equation of second class, one can get

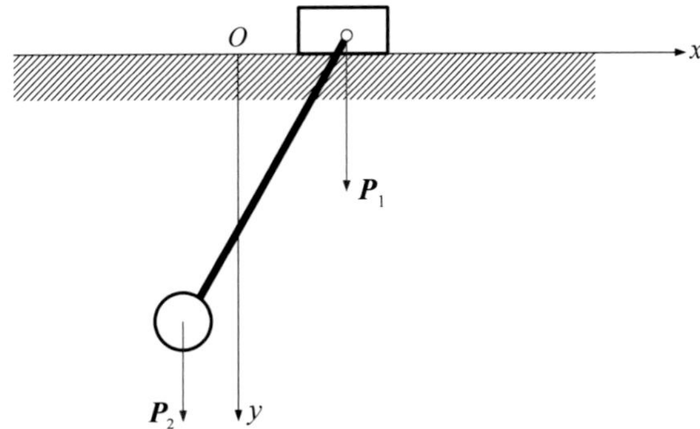

Fig. 14.3 A mass linked with a ball

$$\frac{1}{6}(2m_1 + 9m_2)(R - r)^2\ddot{\varphi} = -M,$$

i.e.,

$$\ddot{\varphi} = \frac{6M}{(2m_1 + 9m_2)(R - r)^2}.$$

Example 3: A mass with gravity $\boldsymbol{P}_1$ is sliding on the smooth level, which is lined with a ball with gravity $\boldsymbol{P}_2$ via a slender bar. The length of the bar is a, and it has no weight. Please build the motion model for this system.

Answer: As shown in Fig. 14.3, the Cartesian coordinate system is built. The length of the bar does not change, and the y-coordinate of the mass is always zero. Thus, the system has two freedom degrees, and we select the angle $q_1 = \varphi$ and $q_2 = x_1$ as the generalized coordinates.

The geometric relations are

$$\begin{aligned} y_1 &= 0, \\ x_2 &= x_1 - a \sin \varphi, \\ y_2 &= a \cos \varphi, \end{aligned}$$

and we have

$$\begin{aligned} v_1 &= \dot{x}_1, \\ v_2^2 &= \dot{x}_2^2 + \dot{y}_2^2 = \dot{x}_1^2 - 2a\dot{x}_1\dot{\varphi}\cos\varphi + a^2\dot{\varphi}^2. \end{aligned}$$

The potential energy at $y_2 = a$ is taken as the zero potential surface, and thus, the potential of the current system is expressed as

$$U = P_2 a(1 - \cos\varphi).$$

The kinetic energy of the system is

$$\begin{aligned} T &= \frac{1}{2}\frac{P_1}{g}v_1^2 + \frac{1}{2}\frac{P_2}{g}v_2^2 \\ &= \frac{P_1 + P_2}{2g}\dot{x}_1^2 + \frac{P_2 a}{2g}\left(a\dot{\varphi}^2 - 2\dot{x}_1\dot{\varphi}\cos\varphi\right). \end{aligned}$$

The Lagrange function is

$$L = \frac{P_1 + P_2}{2g}\dot{x}_1^2 + \frac{P_2 a}{2g}\left(a\dot{\varphi}^2 - 2\dot{x}_1\dot{\varphi}\cos\varphi\right) - P_2 a(1 - \cos\varphi).$$

The Lagrange equation is

$$\begin{aligned} \frac{\mathrm{d}}{\mathrm{d}t}\left(\frac{\partial L}{\partial \dot{q}_1}\right) - \frac{\partial L}{\partial q_1} = 0, \\ \frac{\mathrm{d}}{\mathrm{d}t}\left(\frac{\partial L}{\partial \dot{q}_2}\right) - \frac{\partial L}{\partial q_2} = 0. \end{aligned}$$

We can get

$$\begin{aligned} \frac{\partial L}{\partial \dot{\varphi}} &= \frac{P_2 a}{g}(a\dot{\varphi} - \dot{x}_1\cos\varphi), \\ \frac{\partial L}{\partial \varphi} &= \frac{P_2 a}{g}\dot{x}_1\dot{\varphi}\sin\varphi - P_2 a\sin\varphi, \end{aligned}$$

and thus one can get

$$\frac{P_2 a}{g}(a\ddot{\varphi} - \ddot{x}_1\cos\varphi) = -P_2\sin\varphi.$$

The following equations are:

$$\begin{aligned} \frac{\partial L}{\partial \dot{x}_1} &= \frac{P_1 + P_2}{g}\dot{x}_1 - \frac{P_2 a}{g}\dot{\varphi}\cos\varphi, \\ \frac{\partial L}{\partial x_1} &= 0, \end{aligned}$$

and thus one can get

$$\frac{P_1 + P_2}{g}\ddot{x}_1 - \frac{P_2 a}{g}\ddot{\varphi}\cos\varphi + \frac{P_2 a}{g}\dot{\varphi}^2\sin\varphi = 0.$$

In particular, when the vibration of the speculum is very small, we can have

$$\sin\varphi \approx \varphi,$$
$$\cos\varphi \approx 1,$$
$$\dot{\varphi} \approx 0.$$

These relations can lead to the simplification of the motion equations

$$(P_1 + P_2)\ddot{x}_1 - P_2 a\ddot{\varphi} = 0,$$
$$\ddot{x}_1 - a\ddot{\varphi} - g\varphi = 0.$$

$$\ddot{x}_1 - a\ddot{\varphi} - g\varphi = 0.$$

Removing $\ddot{x}_1$ yields

$$\ddot{\varphi} + \frac{P_1 + P_2}{P_1 a} g\varphi = 0,$$

and its solution is

$$\varphi = A\sin(\omega t + \alpha),$$

where

$$\omega = \sqrt{\frac{P_1 + P_2}{P_1 a} g},$$

and the vibration period is

$$T = \frac{2\pi}{\omega} = 2\pi\sqrt{\frac{P_1 a}{(P_1 + P_2)g}}.$$

When the weight of the mass is much bigger than that of the ball, the vibration period tends to the value

$$T = 2\pi\sqrt{\frac{a}{g}}.$$

Summary

In this chapter, we should master learn the knowledge on Lagrange equation, especially learn how to use it to solve problems in dynamics based on some accurate examples.

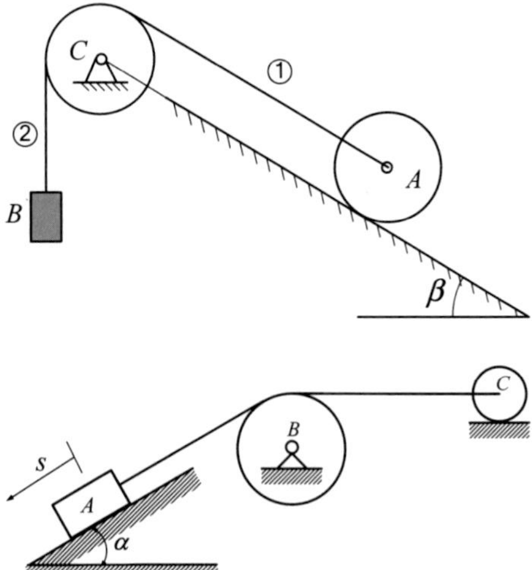

Fig. 14.4 Two wheels and one weight linked together

Fig. 14.5 Two wheels and one weight linked together

Exercises

14.1 As shown in Fig. 14.4, two wheels and one weight are linked by one rope. The wheel A has mass m, and its radius is r, which is in pure rolling on the oblique surface with an angle β. The mass center of the wheel A is linked with a weight with mass $2m$, and the rope is inextensible. The fixed wheel C has the radius r, and its mass is ignored. Please calculate the following variables:

(1) The acceleration $\boldsymbol{a}_A$ of the wheel center A.
(2) The tension $\boldsymbol{T}_1$ of the rope ①.
(3) The frictional force $\boldsymbol{F}$ between the wheel A and the oblique surface.

14.2 As shown in Fig. 14.5, two wheels and one weight are linked together. The gravity of the weight A is $\boldsymbol{P}$, and its dynamic frictional coefficient with the oblique surface is f'. The angle between the oblique surface and the horizontal surface is α, and the weight of the wheel B is $\boldsymbol{P}$ with radius R. There is no relative sliding between the rope and the wheel. The circular disk C is in pure rolling, with the gravity $\boldsymbol{P}$ and radius r. The two segments of the rope are parallel the oblique surface and horizontal surface, respectively. When the weight A falls down along the oblique surface from the stationary state with the distance s, please calculate the following parameters:

(1) The angular velocity and angular acceleration of the wheel B.
(2) The tension force between the weight A and the wheel B.

Answers

14.1 (1) $a_A = \frac{4g-2g\sin\beta}{7}$ (2) $T_1 = \frac{6mg+4mg\sin\beta}{7}$ (3) $f = \frac{(2g-g\sin\beta)m}{14}$

14.2 (1) $\omega_B = \frac{\sqrt{6sg(\sin\alpha-f'\cos\alpha)}}{3R}$, $\varepsilon_B = \frac{g(\sin\alpha-f'\cos\alpha)}{3R}$

(2) $F_{AB} = \frac{2P\left(\sin\alpha-f'\cos\alpha\right)}{3}$.

Chapter 15
Summary

Abstract In this chapter, we will give a brief summary of the course Theoretical Mechanics. The similarity phenomena in nature and analogy method in mechanics are introduced, where all the formulas are grouped and compared, and this is helpful for students to grasp the knowledge in a wider view.

Keywords Analogy phenomenon · Linear quantity · Angular quantity

In fact, it is surprising to see that there exist plenty of similarities and analogy relations throughout the entire chapter of this course. Therefore, we strongly advocate that a new angle, i.e., the analogy relations can be introduced. From this viewpoint, we can find out the internal connections between different phenomena, and then infer the forthcoming law of the second object in comparison with the known characters of the first object. Henceforth, it is beneficial for us to memorize so many contents and grasp the essence of different physical phenomena in so limited time. This new idea also casts a light on how twice as much can be accomplished with half the effort to learn *Theoretical Mechanics*.

Such motivations have stimulated the present chapter, toward trying to explore the possibility of adopting analogy study method to learn *Theoretical Mechanics*. We point out that there are two types of quantities in this class, i.e., the linear quantities and angular quantities.

15.1 Similar Examples in Nature

Maybe we have already noticed a plethora of similar phenomena miraculously created in nature, spanning from nano-, micro-, meso-, macro- even to astronomical scales. Among them, two most well-known examples are the spiral morphologies and hierarchical structures. The nebula, sunflower seed array, sheep horn, fluid vortex, grapevine, snail shell, and even macromolecule all capture the characteristics of spiral or chiral shapes. Some of the typical spiral structures are demonstrated in Fig. 15.1, including the spiral galaxy, water vortex, climbing plant [1], and DNA with double-spiral structure. From the figures, we can see that spiral phenomenon is

J. Liu, *Lecture Notes on Theoretical Mechanics*,
https://doi.org/10.1007/978-981-13-8035-8_15

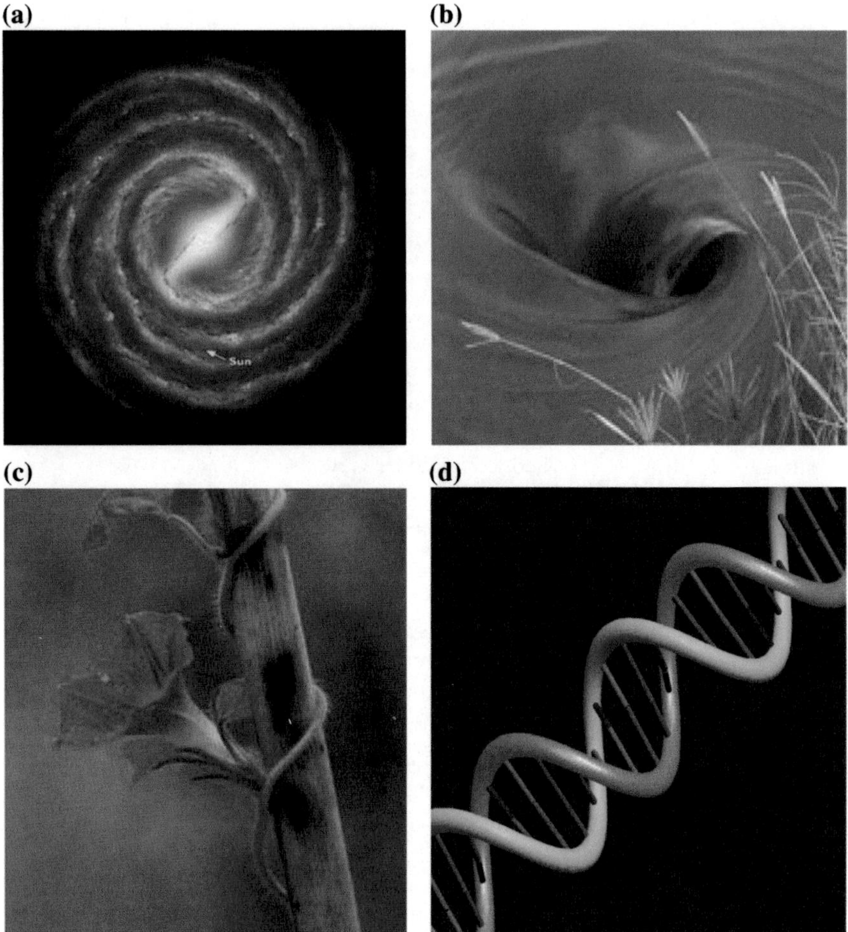

Fig. 15.1 Spiral shapes in nature, **a** spiral galaxy, **b** fluid vortex, **c** climbing plant, **d** DNA with double-spiral structure

ubiquitous in the world, which has been considered as one of the fundamental laws in the world.

Another related phenomenon is the structure hierarchy, which can be observed in the branches and roots of trees, different levels of bone, silk components [2], etc. In addition, the micro/nano hierarchical structure plays a critical role in the self-cleaning capability of lotus [3], superhydrophobicity of water strider leg [4], and strong adhesion force produced by the hairs of gecko feet [5], as shown in Fig. 15.2a–c. Besides these phenomena, if two bundles of hairs are dipped into the liquid, a splendid hierarchical pattern will be formed (as shown in Fig. 15.2d) [6, 7], which is the competition result between the strain energy and surface energy of the liquid.

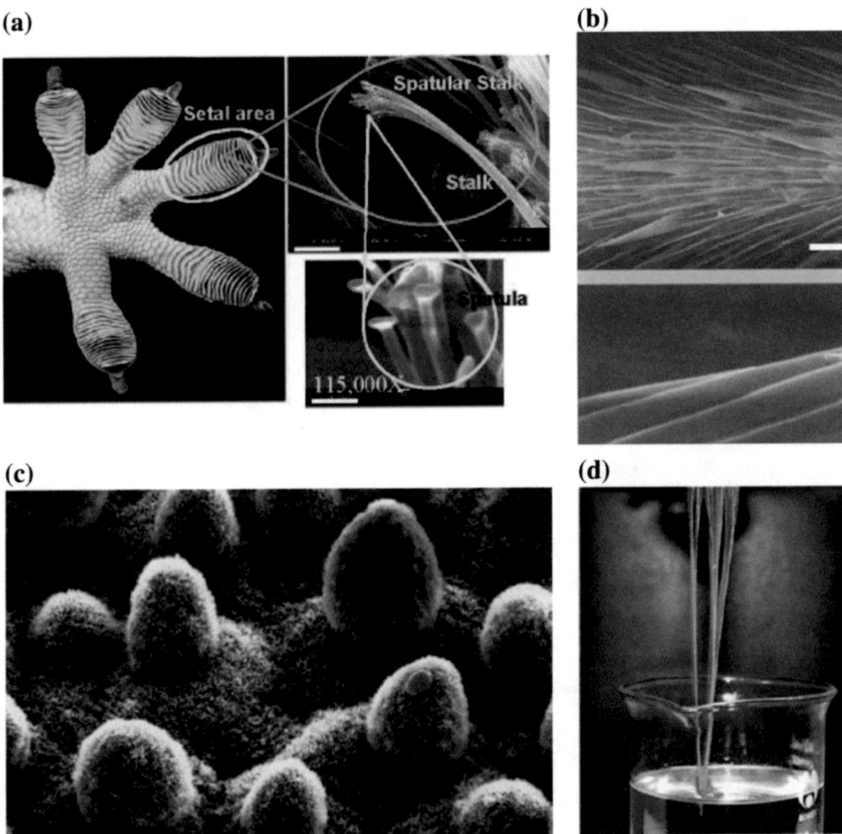

Fig. 15.2 Hierarchical structures in nature, **a** gecko foot, **b** setae of the strider leg, **c** two-level structure of the lotus leaf, **d** hierarchical structure of a bundle of strips dipped into liquid

Moreover, we have already been familiar with some physical laws with similarities in the textbooks, which emerge one after another. For example, a harmonic vibration system and a resistance–inductance–capacitance (RLC) oscillation circuit both obey the same vibration equation. The torsion of a bar can be measured by the membrane analogy or sand-heap analogy for possessing the same format of equation. The course of *Fluid Mechanics* can be taught in comparison with *Theory of Electricity and Magnetism*, for they deal with similar field equations and field quantities. Additionally, a long time ago, the great scientist Maxwell [8] stated that the shape of a meniscus surface is identical with that of an elastic sheet, which was later verified by the experiment by Clanet and Quéré [9], and then was analyzed by Liu [10] in detail.

Besides, much effort has already been devoted to investigating the adhesion similarities between a droplet and a slender rod on a solid substrate. For example, Pugno [11] pointed out that the shape of a carbon nanotube (CNT) cross section resembles the side view of a droplet stuck on a substrate. In succession, Majidi compared three formulations for an elastic adhering to a rigid surface, including stationary principles, surface integral of Eshelby's energy–momentum tensor, and configuration force balance [12]. More recently, Liu and Xia [13] pointed out that the adhesion of a droplet, microbeam, and CNT ring adhered on a solid can all be ascribed into one analysis framework, in which process the strain energy and surface energy compete with each other and reach a final equilibrium state. Although the intrinsic boundary conditions for this sort of problems are fixed, they can be imagined as movable, and the developed analysis framework can be employed successfully. The bending stiffness, characteristic length, energy origination, governing equation, and boundary condition among these three different systems possess close similarities and analogies.

In brief, considering the similarities and analogies existing in different disciplines, the analogy study method can be utilized in learning *Theoretical Mechanics* and *Mechanics of Materials*. This method also paves a new way to design new analogy experiments, and explore the unity of nature in depth.

15.2 Analogies in *Theoretical Mechanics*

Let us then consider the analogies in *Theoretical Mechanics*. The course of *Theoretical Mechanics* mainly includes three sections, namely *Statics*, *Kinematics*, and *Kinetics*. The first part *Statics* is the most rudimentary content in the course of mechanics, and it deals with the simplification of force systems and equilibrium equations. For most students, they have been very familiar with this substance in the middle school stage, so it is not too difficult for them to understand. However, for *Kinematics* and *Kinetics*, it is not so easy for the students as these sections deal with many new concepts and formulas. As a matter of fact, most of the mechanics quantities can be grouped into two classes, i.e., linear quantities and angular quantities. The linear quantities include displacement s (or vector $\boldsymbol{r}$), velocity $\boldsymbol{v}$, accelerator $\boldsymbol{a}$, mass m, force $\boldsymbol{F}$, momentum $m\boldsymbol{v}$, and the corresponding angular quantities are, respectively, angular displacement φ, angular velocity ω, angular accelerator ε, moment of inertia J, moment $\boldsymbol{r} \times \boldsymbol{F}$, and moment of momentum $\boldsymbol{r} \times m\boldsymbol{v}$. The equations for these two classes of quantities are very similar, and they must have some internal connections. Therefore, the analogy relationships of the mechanics quantities are listed in Table 15.1.

In Table 15.1, three kinds of mechanics quantities are demonstrated in detail, i.e., kinetic quantity, inertia quantity, and force quantity. First, from Table 15.1, we can clearly see that the angular quantities including angular displacement, angular velocity, and angular accelerator, are in lower orders than the corresponding linear quantities. Second, if two mechanics quantities are properly combined, new mechanics quantities, such as force $\boldsymbol{F} = m\boldsymbol{a}$, moment $\boldsymbol{M} = \boldsymbol{r} \times \boldsymbol{F}$, momentum $\boldsymbol{P} = m\boldsymbol{v}$, impulse $\boldsymbol{I} = \boldsymbol{F}t$, moment of momentum $\boldsymbol{L}_o = \boldsymbol{r} \times m\boldsymbol{v}$, kinetic energy $T = \frac{1}{2}mv^2$, and work $W = Fs$ can be produced. Moreover, the most important theorems in Table 15.1 are Newton's second law, theorem of momentum, theorem of moment on momentum,

Table 15.1 Linear and angular quantities in *theoretical mechanics*

Mechanics quantity	Linear quantity	Angular quantity	Connection
Displacement	$\boldsymbol{r} = \boldsymbol{r}(t)$ $s = s(t)$	$\varphi = \varphi(t)$	$s = \varphi r$
Velocity	$\boldsymbol{v} = \dot{\boldsymbol{r}}$ $v = \dot{s}$	$\omega = \dot{\varphi}$	$v = \omega r$
Accelerator	$\boldsymbol{a} = \dot{\boldsymbol{v}} = \ddot{\boldsymbol{r}}$ $a_\tau = \dot{v} = \ddot{s}$	$\varepsilon = \dot{\omega} = \ddot{\varphi}$	$a_\tau = \varepsilon r$
Inertia quantity	m	J	$J = \sum m_i r_i^2$
Force	$\boldsymbol{F} = m\boldsymbol{a}$	$M = J\varepsilon$	$\boldsymbol{M}_o(\boldsymbol{F}) = \boldsymbol{r} \times \boldsymbol{F}$
Momentum	$\boldsymbol{P} = m\boldsymbol{v}$	$L_0 = J\omega$	$\boldsymbol{L}_o = \boldsymbol{M}_o\,(m\boldsymbol{v}) = \boldsymbol{r} \times (m\boldsymbol{v})$
Theorem of momentum	$\frac{\mathrm{d}\boldsymbol{P}}{\mathrm{d}t} = \sum \boldsymbol{F}$	$\frac{\mathrm{d}\boldsymbol{L}_o}{\mathrm{d}t} = \sum \boldsymbol{M}$	
Kinetic energy	$T = \frac{1}{2}mv^2$	$T = \frac{1}{2}J\omega^2$	$\mathrm{d}T = \mathrm{d}W$ $T_2 - T_1 = W$
Work	$T = \int_C F\mathrm{d}s$	$W = \int_\varphi M\mathrm{d}\varphi$	

and theorem of kinetic energy. For the rigid body, Newton's second law corresponds to the theorem of momentum, and degenerates into the mass center motion equation of a rigid body

$$m\boldsymbol{a}_C = \sum \boldsymbol{F}_i,$$

where $\boldsymbol{a}_C$ is the accelerator of the mass center of the rigid body and $\sum \boldsymbol{F}_i$ is its resultant force. Similarly, the theorem of moment on momentum can be reduced to the differential equation of a rigid body's motion with a fixed axis

$$J_z\varepsilon = \sum M_i,$$

where J_z is the inertia of moment for the rigid body on a fixed z-axis and $\sum M_i$ is the resultant moment.

Altogether, if the students can grasp these relations, they surely can have a bird's view on the whole substance of *Theoretical Mechanics*, and accordingly, they can amount to the effect of showing all the contents in one paper only.

In conclusion, the course of *Theoretical Mechanics* mainly deals with two types of quantities, i.e., the linear quantities and angular quantities, and these concepts and equations have similar analogy relations. As a consequence, through the full explorations of all kinds of analogy relations in *Theoretical Mechanics*, it is found that there are really a lot of formulas with similarities. In use of this analogy study method, we can easily master these concepts and formulas. It is beneficial for us to have a bird's view on the mechanics knowledge, and then twice as much can be accomplished with half the effort. This new study method also opens a new path to design new-typed analogy experiments, and gain a full insight into the unity of nature.

References

1. Goriely A, Tabor M. Spontaneous helix hand reversal and tendril perversion in climbing plants. Phys Rev Lett. 1998;80(2):1564–7.
2. Zhao HP, Feng XQ, Cui WZ, Zou FZ. Mechanical properties of silkworm cocoon pelades. Eng Frac Mech. 2007;74(12):1953–62.
3. Neinhuis C, Barthlott W. Characterization and distribution of water-repellent, self-cleaning plant surfaces. Ann Bot–London, 1997; 79(6): 667–77.
4. Hu DL, Chan B, Bush JW. The hydrodynamics of water strider locomotion. Nature, 2004; 424(6949):663–6.
5. Gao HJ, Wang X, Yao HM, Gorb S, Arzt E. Mechanics of hierarchical adhesion structure of gecko. Mech Mater. 2005;37(2–3):275–85.
6. Bico J, Roman B, Moulin L, Boundaoud A. Adhesion: elastocapillary coalescence in wet hair. Nature. 2004;432(7018):690.
7. Liu JL, Feng XQ, Xia R, Zhao HP. Hierarchical capillary adhesion of micro-cantilevers or hairs. J Phys D-Appl Phys. 2007;40(18):5564–70.
8. Maxwell JC. Capillary action. IN: Encyclopaedia Britannica, vol. II. London: Cambridge University Press; 1980.
9. Clanet C, Quéré D. Onset of menisci. J Fluid Mech. 2002;460(1):131–49.
10. Liu JL. Analogies between a meniscus and a cantilever. Chin Phys Lett. 2009;26(11):116803.
11. Pugno NM. An analogy between the adhesion of liquid drops and single walled nanotubes. Scripta Mater. 2008;58(1):73–5.
12. Majidi C. Remarks on formulating an adhesion problem using Euler's elastica (draft). Mech Res Commun. 2006;34(1):85–90.
13. Liu JL, Xia R. A unified analysis of a micro-beam, droplet and CNT ring adhered on a substrate: variation with movable boundary condition. Acta Mech Sin. 2013;29(1):62–72.

Printed in the United States
By Bookmasters